FLOOD RISK MANAGEMENT

Our changing climate and more extreme weather events have dramatically increased the number and severity of floods across the world. Demonstrating the diversity of global flood risk management (FRM), this volume covers a range of topics including planning and policy, risk governance and communication, forecasting and warning, and economics. Through short case studies, the range of international examples from North America, Europe, Asia and Africa provide analysis of FRM efforts, processes and issues from human, governance and policy implementation perspectives. Written by an international set of authors, this collection of chapters and case studies will allow the reader to see how floods and flood risk management is experienced in different regions of the world. The way in which institutions manage flood risk is discussed, introducing the notions of realities and social constructions when it comes to risk management.

The book will be of great interest to students and professionals of flood, coastal, river and natural hazard management, as well as risk analysis and insurance, demonstrating multiple academic frameworks of analysis and their utility and drawbacks when applied to real-life FRM contexts.

Edmund C. Penning-Rowsell is Professor of Geography and Pro Vice-Chancellor at Middlesex University, where he founded the Flood Hazard Research Centre in 1970. Since 2010 he has been a Visiting Academic at the School of Geography and the Environment at the University of Oxford. He is also the editor of the journal *Environmental Hazards* (Taylor & Francis).

Matilda Becker is a DPhil researcher at the School of Geography and Environment, University of Oxford. Her research centres on legal geographies of mineral exploration in the Canadian Arctic. Matilda previously worked as a research assistant on a public engagement project for flood risk management in Yorkshire, England. She graduated from the University of Oxford with an MSc in Water Science, Policy and Management.

EARTHSCAN WATER TEXT

For more information about this series, please visit: https://www.routledge.com/
Earthscan-Water-Text/book-series/ECEWT

Water Governance and Collective Action
Multi-scale Challenges
Edited by Diana Suhardiman, Alan Nicol and Everisto Mapedza

Water Stewardship and Business Value
Creating Abundance from Scarcity
William Sarni and David Grant

Equality in Water and Sanitation Services
Edited by Oliver Cumming and Thomas Slaymaker

Environmental Health Engineering in the Tropics
Water, Sanitation and Disease Control
(third edition)
Sandy Cairncross and Sir Richard Feachem

Flood Risk Management
Global Case Studies of Governance, Policy and Communities
Edited by Edmund C. Penning-Rowsell and Matilda Becker

Water Ethics
A Values Approach to Solving the Water Crisis
(second edition)
David Groenfeldt

FLOOD RISK MANAGEMENT

Global Case Studies of Governance, Policy and Communities

Edited by Edmund C. Penning-Rowsell and Matilda Becker

Routledge
Taylor & Francis Group

LONDON AND NEW YORK

earthscan
from Routledge

First published 2019
by Routledge
2 Park Square, Milton Park, Abingdon, Oxon OX14 4RN

and by Routledge
52 Vanderbilt Avenue, New York, NY 10017

Routledge is an imprint of the Taylor & Francis Group, an informa business

Library of Congress Cataloging-in-Publication Data
Names: Penning-Rowsell, Edmund, editor. | Becker, Matilda, editor.
Title: Flood risk management : case studies of governance, policy and
communities / edited by Edmund Penning-Rowsell and Matilda Becker.
Other titles: Flood risk management (Routledge)
Description: New York, NY : Routledge, 2019. |
Includes bibliographical references.
Identifiers: LCCN 2018043435| ISBN 9781138541900 (hardback) |
ISBN 9781138541917 (pbk.) | ISBN 9781351010009 (e-book)
Subjects: LCSH: Flood damage prevention--Case studies. |
Floodplain management--Case studies. | Risk assessment--Case studies.
Classification: LCC TC530 .F583 2019 | DDC 363.34/936--dc23
LC record available at https://lccn.loc.gov/2018043435

ISBN: 978-1-138-54190-0 (hbk)
ISBN: 978-1-138-54191-7 (pbk)
ISBN: 978-1-351-01000-9 (ebk)

Typeset in Bembo
by Taylor & Francis Books

Printed and bound in Great Britain by
TJ International Ltd, Padstow, Cornwall

CONTENTS

List of illustrations   *vii*
List of contributors   *ix*
Foreword   *xiii*

INTRODUCTION

1 Realities and social constructions in flood risk management   1
   Edmund C. Penning-Rowsell and Matilda Becker

A. GOVERNANCE

2 Legal geography and flood risk management in Germany   17
   Matilda Becker

3 The changing nature of financing flood damages in Canada   30
   Heather Bond

4 Power for change in adapting to coastal flood risk on Curacao
   in the Caribbean   43
   Lena Fuldauer

5 Power shifts in flood risk management: Insights from Italy   58
   Andrea Farcomeni

6 'Going Dutch' in flood risk management: How is Dutch flood
   policy mobilised?   69
   Timo Maas

B. POLICY AND IMPLEMENTATION

7 Flood policy process in Jakarta, Indonesia, using the Multiple
   Streams model   79
   Thanti Octavianti

8 A revolving door of policy evolution: Climate change
   adaptation after Superstorm Sandy   91
   Carey Goldman

9 Policy belief change and learning in response to California
   flooding   103
   Clarke A. Knight

10 The challenges of flood warning systems in the developing
   world   115
   Mahala McLindin

11 Flood warning and recovery in Zimbabwe: Some salutary
   lessons   131
   Abigail Tevera

C. COMMUNITY AND PEOPLE

12 Adapting to floods in social housing in the UK: A social justice
   issue   141
   Diana King and Edmund C. Penning-Rowsell

13 Emergency intentional flooding: Is social justice adequately
   considered?   153
   Anne Muter

14 At the water's edge: Motivations for floodplain occupation   165
   Laura West Fischer

15 Flood insurance maps and the US National Flood Insurance
   Program: A case for co-production?   177
   Allison Reilly

16 The effectiveness of social media in flood risk communication   188
   Wenhui Wu

Index   *202*

ILLUSTRATIONS

Figures

1.1	The complexity of interventions to reduce flood risk	11
4.1	Steps in the hybrid methodology developed in this research: multi-level power dynamics mapping (MLPDM)	45
4.2	Multi-stakeholder influence map of Curacao's FRM regime in 2017, following the method of Sova et al. (2015)	48
4.3	Multi-level power dynamics map showing power relationships between most influential actors in the FRM regime in Curacao, 2017	49
10.1	The 'chain' in flood early warning systems (adapted from Parker, 2003).	117
11.1	The flood forecasting, warning and response chain and key supporting processes	132
16.1	Risk communication throughout the risk management cycle (after Höppner et al., 2012)	189

Tables

2.1	Insurance coverage in Germany by Land. Länder in bold were previously part of the German Democratic Republic (East Germany). Data adapted from GDV, 2018.	22
3.1	Level of consensus among insurance companies interviewed, concerning key risks and opportunities with introducing overland flood insurance (modified from Thistlethwaite and Feltmate, 2013, 38).	36

3.2   Summarised comparison of international flood insurance
      programmes of G8 countries, excluding Canada (modified
      from IBC, 2015, 13)   38
4.1   Overview of selected power theories used in this chapter   44
4.2   Interviewees (conducted in December 2017)   46
5.1   Negotiation and power distribution in different flood policy
      epochs in Italy   61
6.1   Interviewee organisations   72
7.1   Key activities and events related to Jakarta's flood policy
      evolution since the mid-1990s   82
7.2   Key characteristics of the Multiple Streams model   87
8.1   Interview subjects and questions   93
9.1   Stated DWR goals with corresponding deep core beliefs and
      policy core beliefs   107
9.2   A summary of stated DWR missions, key actors involved in
      water plan updates, and the corresponding policy core
      beliefs (taken from the DWR water plans 1987–2018)   110
11.1 Key informant interviewees   134
12.1 The lack of take-up of flood insurance by householders in
      different tenure/housing types   148
14.1 Factors that influence the decision to live on a floodplain 170
15.1 Potential barriers to knowledge production within the Risk
      MAP process   181
16.1 The use of social media during natural disasters including
      flooding   190

CONTRIBUTORS

Matilda Becker is a DPhil researcher at the School of Geography and Environment, University of Oxford. Her research centres on legal geographies of mineral exploration in the Canadian Arctic. Matilda previously worked as a research assistant on a public engagement project for flood risk management in Yorkshire, England. She graduated from the University of Oxford with an MSc in Water Science, Policy and Management. Outside academia, Matilda enjoys attending debating clubs, swimming and horse riding.

Heather Bond is a Project Officer at the International Water Resources Association in Paris, France. She works on projects relating to water quality and smart water management, including directing a task force of global experts and organising sessions at international events, with a specialisation in flood management and wetlands. She previously worked for Environment Canada and the Ontario Ministry of the Environment in landscape assessment of aquatic habitats. In her spare time, Heather enjoys travelling, yoga and playing the piano.

Andrea Farcomeni currently works as Senior Asset Engineer at Affinity Water. He has contributed to the preparation and submission of the Water Resources Management Plan 2019, leading the technical work on demand forecast and investment appraisal. Andrea graduated from the University of Oxford with an MSc in Water Science, Policy and Management. In his spare time he enjoys hiking and travelling, but above all he loves cooking.

Lena Fuldauer is starting her PhD at the Environmental Change Institute at the University of Oxford. Her research specialises in sustainability transitions, decision-making and the role of resilient infrastructure networks for the most vulnerable countries to climate change. Previously, she worked for the United Nations Office for Project Services as a technical assistant, developing tools for evidence-based infrastructure assessments and long-term strategic planning for sustainable development outcomes. In her free time she loves exploring new adventures, hiking up mountains and cooking creatively.

Carey Goldman currently manages trainings and communications for Clover Food Lab, in Boston, MA, working to create a more sustainable food service industry with a reduced carbon footprint. Previously, Carey served as a Conservation Trust of North Carolina AmeriCorps member with the University of North Carolina Wilmington's MarineQuest programme. Outside work you can find him on, near or underneath the water sailing, scuba diving or simply enjoying a good book on the beach.

Diana King works for the UN World Food Programme as a Policy Programme Officer on the linkages between humanitarian and development work in Malawi. Previously, she worked as a research consultant at the Overseas Development Institute and as a research assistant for the Climate Change Adaptation Research Group at McGill University, Canada. Diana holds an MSc in Environmental Change and Management from the University of Oxford. She enjoys learning languages and cooking in her spare time.

Clarke Knight is currently a PhD candidate in Environmental Science, Policy and Management at UC Berkeley. She conducts research on the carbon sequestration potential of California's mixed conifer forests. Before starting her PhD, Clarke received two master's degrees – an MSc in Water Policy and an MSc in Biodiversity Conservation – from the University of Oxford where she studied as a Rhodes Scholar, and a BA in Chemistry from Smith College

Timo Maas currently works as Researcher at the Rathenau Instituut in The Hague, the Netherlands, which specialises in the social aspects of science, technology and innovation. Timo's research focuses on innovation, examining new forms of collaboration and emerging policy paradigms. Before joining the Rathenau Instituut, Timo worked as a researcher of sustainable development at Kaleidos Research/NCDO. When not at work, Timo can be found on a tennis court or in the cinema – depending on the weather.

Mahala McLindin is a water resources engineer with experience in water resource modelling, design, management and strategic planning. She has worked on sustainable and integrated planning of river, water supply, wastewater and storm water systems in Australia and has conducted research on global water challenges. Mahala now works for the New South Wales Government in water trading within the Murray Darling Basin. She has a degree in Civil Engineering and a MSc in Water Science, Policy and Management. Mahala enjoys hiking, skiing and snorkelling.

Anne Muter is a lawyer in Vancouver doing a combination of general civil litigation and water law work. She has consulted on transboundary water issues as well as groundwater and environmental flow policy. Anne has a master's degree in Water Science, Policy and Management in addition to a law degree and an undergraduate degree in Oceanography and Atmospheric sciences. She enjoys getting out on to the water in a sailboat, canoe or for a swim as often as she can.

Abigail Tevera currently works as a WASH Officer at United Nations Children's Fund (UNICEF) in Harare, Zimbabwe, providing technical, operational and administrative assistance throughout the Water, Sanitation and Hygiene (WASH) programming process. Prior to joining UNICEF, Abigail has around 18 years of work experience in International Humanitarian and Development – on missions in Zimbabwe, Liberia and Yemen, including with the NGO Oxfam. Outside of her professional life she enjoys bonding with nature through gardening and farming activities. She also sings in a local church choir and networks with a variety of business fraternity.

Thanti Octavianti is a final-year DPhil researcher at the School of Geography and the Environment, University of Oxford. She investigates the political dynamics of a large water infrastructure project to enhance urban water security, using Jakarta's seawall megaproject as a case study. She holds an MSc in Water Science, Policy and Management from Oxford and a BEng in environmental engineering from Universitas Indonesia. Prior to coming to Oxford she worked as a consultant in the Ministry of Public Works, Indonesia, assessing local governments' performance in the provision of water services. Outside work, Thanti enjoys doing some 'experiments' in her kitchen.

Edmund C. Penning-Rowsell is Professor of Geography and Pro Vice-Chancellor at Middlesex University, where he founded the Flood Hazard Research Centre in 1970. Since 2010 he has been a Visiting Academic at the School of Geography and the Environment at the University of Oxford. His extra-curricular interests include (vegetable) gardening and wine.

Laura West Fischer currently works as an Engineer/Scientist at the Electric Power Research Institute (EPRI) in Washington, DC. Her research specialises in resilience, climate impacts, and integrated assessment as they relate to the electric power sector. Prior to EPRI, Laura worked in climate change adaptation at the US Environmental Protection Agency and disaster services at the American Red Cross of Alaska. Outside of her professional life, she is an avid swimmer and board game enthusiast.

Allison Reilly currently works as an Associate at WSP US Advisory Services in Washington, DC. In her role, she supports an array of projects focused on developing, managing and improving transportation and infrastructure systems across the

US. Prior to joining WSP, Allie completed her MSc at the University of Oxford's School of Geography and the Environment, focusing her research on the dynamic relationship between urban transit and citizenship. In her free time she enjoys playing football and patronising neighbourhood bookstores.

Wenhui Wu has been working as a Senior Officer at the British Consulate-General Guangzhou (BCG) in China for the last three years, focusing on promoting collaboration and exchanges between China and the UK to drive green developments in South China. Before taking on her current role she worked for the Foreign and Commonwealth Office as a Sustainable Urbanisation Policy Officer at BCG. Wenhui is an animal lover and environmental enthusiast. Having lived in China, the USA, the UK and Australia, she likes to explore cultural differences and enjoys playing tennis.

FOREWORD

Change is afoot in the world of flood risk management: warnings of floods are communicated instantly around the world; satellites map flood extent in every floodplain on earth; global flood models predict flood risk in places that the modellers have never visited. Yet people who experience flooding often do not recognise that every flood and every place is different. Yet there are similarities: the ways floods materialise in practice are shaped by local geography and the institutions that have responsibility for managing flood risks and assisting recovery. Each place and each institution bears the hallmarks of local culture and politics. Floods and these similarities cannot be understood without understanding their local particularities.

The chapters in this book illustrate the importance and diversity of context. Looking through a variety of different lenses, we see how floods are experienced by people in different places and how fallible institutions shape the ways in which flood risk is managed. The narratives introduce notions of power, information, knowledge, truth and some challenges relating to flood risk management, and emphasise the attention that we must pay to what is considered 'real' and what is framed as a 'construction'.

These chapters have emerged from a network of researchers associated with the School of Geography and the Environment in the University of Oxford, marshalled by Professor Edmund Penning-Rowsell in partnership with Matilda Becker. Only in a university as international as Oxford could there be such a diversity of nationalities and scholarly commitment to travel near and far, to scrutinise the lived experience of flooding though the various theoretical lenses that they acquired amongst the dreaming spires. This book therefore covers a remarkable geographical sweep, taking examples of flood risk management from Canada to Zimbabwe, in developed and in developing countries. The different chapters use different social science ideas and theories, drawing from geography, sociology, political science and elements of psychology.

This concentration of case studies is therefore uniquely valuable. Such examples of the reality of making decisions about flood risk emphasise the complexity of such processes and the importance of finding sustainable solutions. The multiplicity of the methods used in the research leading to these chapters will help future students of the subject choose an appropriate methodological path towards insightful policy prescriptions.

Jim W. Hall FREng | Professor of Climate and Environmental Risks | Director of the Environmental Change Institute, University of Oxford

1

REALITIES AND SOCIAL CONSTRUCTIONS IN FLOOD RISK MANAGEMENT

Edmund C. Penning-Rowsell and Matilda Becker

Introduction

Floods, and tackling or managing flood risk, are about realities. A family finding 60 cm of dirty floodwater in their dwelling is an inescapable reality. A flood embankment or levee that can be seen and kicked is also a reality, even if its construction is a product of social processes and political negotiations. A company offering flood insurance is the same. The warning of impending flooding is also a social product in that definitions of risk and our responses to them are parameterised and defined in relation to the flood hazard as perceived.

Consider a thunderstorm event or a hurricane approaching: a forecast based on meteorological models making assumptions about that storm and a warning message accompanying it are both social constructions in that they are an interpretation of what might happen and what damage and disruption could be avoided.

Consider also the idea that a series of floods over several years in a single region generate different and socially contingent discourses of causality, uncertainty, management responsibilities and political implications. The hazard plays out against a colourful and unstable backdrop, flavoured by the socio-environmental politics of the day. This backdrop affects how we interpret and re-interpret the hazard that flooding presents and how we position ourselves as sections of society in relation to those floodwaters.

Dig a little deeper and realise that our social relation to the hazard is influenced by different agents and their power, what we consider to be truth, what information we trust and what uncertainties we recognise – and ultimately how these interact to create challenges and opportunities for us to manage flood risk and successfully live with flood events.

In this context, this chapter introduces notions of power, information, knowledge, truth, and some challenges relating to flood risk management (FRM), and emphasises the attention which we must pay to what may be considered 'real', what is framed as a 'construction', and by whom.

So, any analysis of FRM, as in this volume, is a blend of objective facts and subjective interpretations, and in this respect we take a position sometimes described as 'weak constructivism' (Lupton, 1999). To clarify, *social constructionism* focuses on the societal scale (the macro), whereas *constructivism* focuses on the individual within the societal context (the micro). There is, of course, a two-way flow between these scales.

However, whatever it is called, we wish to stress in this volume that the management of risks is a social process, and indeed that the risks as we see them are a mixture of fact and supposition. Governance arrangements are product of history and social processes, and policies are normative statements of intentions that are heavily influenced by the society from which they come. Communities of people react to realities that they observe but they also have interpretations that are conditioned by the risk that they face and their community's character, its history and their aspirations.

Agents and power in flood risk management

The issue and challenges

The products of FRM, in general, are determined by those who have power. Theoretical conceptualisations of power are multiple and diverse, however all consider the relations between different actors (see Weber (1964); Foucault (1979); Allen (2009); Ailon (2006); and Lukes (2004)). One application of models of power to the field of FRM is provided by Self and Penning-Rowsell (2017).

Power might enable one set of actors to influence others to act in a certain way, whether they oblige or not (Weber, 1964). Pfohl (2008) considers the Weberian approach important for understanding how society is hierarchised, and thus how social relations are constructed. Pfohl (2008, 12) adds to Weber's definition of power in three ways: by considering power as more than just a resource held by one set of actors (i.e. that power works by inclusion and exclusion, not just by force); by differentiating between coercive and hegemonic power; and the interaction between the 'global coloniality of power'. This latter point is significant to social constructionism, as it highlights that power is a relation which is continually remade, reasserted and redefined. It is not stable. Foucault's (1979) work develops this further. Lastly, and most significantly for the framing of the remainder of this chapter, Foucault (1979) argues that power through discourse (i.e. power to maintain a dominant paradigm of thought and create histories based upon it) is critical in sustaining and producing social relations.

So, speaking in *realist* terms, what is power in this type of field? It includes the power to raise and spend money, to provide services (often regulated), to say 'yes' or 'no' to proposals or plans, and the power to set the overall general policy direction (Birkland, 2016). It is also the power to investigate and 'punish', to direct people and organisations, and the power to influence ideas and debates in this and related fields.

Power in flood risk management

Within FRM, the power to raise and spend money is important because, in general, much risk reduction is a capital-intensive activity where large sums are needed to implement significant flood warning systems and for protecting local communities and regional economies (chapter 7 herein). Insurers need very substantial financial resources to ensure they can cover everything possible in the claims they receive and remain solvent (chapter 3). In general, governments control the raising of funds in this way, but these funds may be raised by regional organisations, or at catchment or coastal level, or within the city boundaries. The power to provide services that are regulated properly by government organisations includes the power to sell insurance cover within private or regulated markets. This power is often constrained by regulatory provisions, set by government or quasi-government organisations, safeguarding the solvency of insurance schemes or companies and protecting the public from their collapse.

Flood risk management is also often driven by actors in the form of specialist agencies (in either the water field or the environmental protection area) or disaster management authorities. These organisations may have legal powers, allowing them to operate in constructing flood defence schemes or providing services, with these powers often delegated from national governments and steered by national legislation (chapters 2 and 15). Science-based organisations such as national meteorological and insurance departments also have the power that knowledge brings (chapter 16), particularly in our case concerning risk levels both long term and immediate.

The power to say 'yes' or 'no' in FRM is best exemplified by local spatial planning organisations having the power to regulate the use of land at risk from flooding, and prevent the build-up of current and future flood damage potential (chapter 14). These powers may be circumscribed in most countries by some geographical delineation of the at-risk areas (i.e. the flood plain), but such powers are often a small part of an overall land use planning system designed to zone all areas for particular uses and make decisions on specific development applications.

When tackling a flood event, the 'blue light' organisations (the police; the ambulance service; coastguards; the fire brigade; etc.) generally have the power to direct people –for example, to evacuate their properties (chapter 10). Non-governmental relief organisations such as the Red Cross and Red Crescent may have similar powers (chapter 4). They may, for example, have the power to set up encampments for those displaced by floods or use the power of other organisations to the same effect (chapter 11).

The field of FRM is continually changing, driven by international, national and regional debates about the scale of risk that communities face and the measures necessary to implement risk reduction. Foucault would consider these 'debates' as shifting movements of power in discourse. In Britain, for example, we have seen the debate change over the last four to five decades from land drainage for agricultural intensification, through to flood defence, through to FRM, reflecting different philosophies about the relationship between nature and societies (Tunstall et al., 2004).

Debates in other countries have followed similar directions, driven often by a recognition that an approach to reducing flood risk dominated by engineering measures, in the face of rising risk levels, is neither sustainable nor affordable (chapters 6 and 9). Social constructions underpin such recognitions, particularly in the parameterisation of what is 'affordable', 'sustainable' and 'valuable'. The agents articulating such debates include science organisations, researchers and professional bodies (chapter 6), and each has some power to influence the agendas of government and non-government organisations, although often this power and its effects are hidden from view.

Finally, as the FRM field matures (and the role of specialist agencies such as the Environment Agency in England and the US Army Corps of Engineers in United States diminishes somewhat), it is clear that communities have some power, not least the power to accept or reject proposals by these and other specialist agencies or by government. But this raises the question as to whether this power is *enough* for communities or whether it might even be *disruptive* to effective management. The answer here depends on what you see the role of citizens in hazard management to be – and indeed your personal and community experience of such issues.

The challenge within FRM today is to recognise the disparate nature of the power structures involved and for governments and others to make progress by using these powers to maximise risk reduction for those at risk. This is no easy matter. Floods are episodic events and, for example, what may appear to be sensible powers during flood emergencies may appear unsatisfactory and overbearing in flood-free periods (chapter 5). The integration of FRM proposals within water planning or disaster management may prove difficult if FRM powers are too precisely prescribed – so as to properly fit that sector – rather than designed to assist integration with others.

Power and agency exemplified

Other ideas on power are deployed in chapter 4 for the island of Curaçao in the Caribbean (Dahl, 1957; Emerson, 1962; Bachrach and Baratz, 1962; Lukes, 2004). Here the most powerful are agencies concerned with safety issues and weather forecasting but the tourism industry – with its hotels located (and to be located) on the coast – exerts more influence here than elsewhere; surprisingly, the influence on investment of the World Bank and United Nations agencies has not been significant.

A broader analysis of power is provided in chapter 5 for Italy where the power dynamics in FRM, as in other countries (e.g. in chapter 14), are the result of the constant interplay between central and local forces. The different agents involved in FRM at different scales in Germany – at federal and regional level – are shown in chapter 2. The importance of social media as powerful agents of risk communication – creating new social constructions of risk and relations to the hazard – are evident in chapter 16. Those with power and influence can steer policy change (e.g. Johnson et al., 2005), and in the post-disaster situation in New York (chapter 8) policy change arrived because of the interplay of national and state agents, but with local communities determining much of the post-disaster discourse. The power of stakeholder groups to affect government policy towards insurance in Canada is examined in chapter 3.

Information and knowledge

'Knowledge is power' is often quoted from Francis Bacon in the 1500s. However, building on the theoretical bases of power in the section above, we examine here how only knowledge that is considered to contain the *right* information has power to influence. This section charts how the *right* information can be represented *reliably*, and how this reveals patterns and challenges in information *flows*.

When implementing FRM, information and knowledge provide a foundation for understanding, preparing for and reacting to flood risk. Significant differences may be found in the types and sources of information which inform people's knowledge and perception of this risk and which information they choose to trust. These differences produce both challenges and opportunities for enhancing the efficacy of FRM. Challenges arise where information is unreliable, unusable or carries high levels of uncertainty. FRM in recent decades has demonstrated that trust is a two-way process, with both scientists and lay publics often having reservations about the validity of, or indeed motivations behind, the flood risk information and knowledge provided (Faulkner et al., 2007).

An important issue can be identified here: that of normativity (what *ought* to be) and positivity (what *is*). It is helpful to think of knowledge within these two categories because it uncovers ways of thinking in socio-political and scientific-technical approaches to FRM. There is considerable overlap between these two concepts and realism and constructionism. Bankhoff et al. (2004) explain that different constructions of normativity and positivity affect both quantitative and qualitative sciences alike. Applied scientists – for example, hydrologists and economists – may claim to take a realist or positivist approach to a hazard because they are able to quantify it. However, the quantification of a hazard (how much rain falls and flows down rivers) is not equal to the parameterisation of its associated risks. The latter are social products of the human-ecological system and are therefore normative social constructions of risk and reality.

Scientific/technological normativity

Flood risk modelling provides an excellent example of what a system ought to be like from the perspective of an economist modelling potential losses. Within flood management, normativity (probabilistic modelling) is predominantly used to model flood risk, scale, recurrence and hence likely economic impacts. Deterministic statements (based on positive knowledge) underpin several of the models which might be used to produce different flood maps of floodplain areas and hence, again, potential future flood impacts. For example, a model may work on the positivist statement that water will flow downhill, and on to the floodplain, but also normative ones such as 'at depth X, probability Y, and duration Z, affecting properties A to N, then flood water *ought* be the cause of so much damage and disruption affecting R, S and T communities'. Different components of the equation, particularly in relation to flood duration, contain variegated levels of uncertainty affecting other components in the equation and the model more generally (Penning-Rowsell et al., 2013). Therefore, the output of such a model will give a range of possibilities, accounting for uncertainty relating to different elements of the flood loss generating system.

An effect of this, given that such economic loss modelling is never absolute in relation to floods, is the challenge of communicating uncertainty, probability and risk to non-expert audiences. These audiences include not only the public but also practitioners in policy and front-line FRM agencies (Faulkner et al., 2007). Despite the uncertainty of probabilistic models, their outputs must be communicated reliably and effectively to engender trust from their users in the information they provide. Demeritt et al. (2013), for example, give the perspectives of information users for improving flood-warning communication. Beven (2016) provides suggestions for improving early translations of uncertainty between model and practitioner. In this volume, the impact of broken or ineffective communication systems is explored in both chapters 10 and 11, wherein the devastating impacts of poor warning systems are outlined and technical flood risk information is demonstrated to be difficult to reconcile with diverse, non-expert knowledge communities, particularly in the developing world.

Socio-political normativity

Significant theoretical literature can be found on normativity and positivity in environmental management. Jasanoff (2004) describes how knowledge is never pure but is supported and informed by social norms, moralities and often political motivations. Evidence of this can be found in the manner in which epistemologies are differentially treated by both 'expert' and 'lay' people. Using a science and technology studies framework, Jasanoff (2004) emphasises how normativity affects how we interact with people, objects and space. A more general argument can be found in Foucault's (1972) epistemes, a theory concerning our unconscious biases towards certain knowledges and information histories that we assume to be true based on socio-cultural assumptions or learning.

This is starkly demonstrated in chapter 6, where the expropriation of Dutch flood management culture and approaches to developing countries demonstrates something of an impasse between local and 'imported' knowledge. Issues of power,

knowledge and risk perception are understood as differential and socially constructed against a historical-environmental backdrop of flood management.
Additionally, chapter 9 considers how normative, culturally underpinned attitudes towards nature, water, the environment and engineering have impacted the
FRM approach taken by the Californian government. The chapter demonstrates
how biases towards certain knowledge histories and trajectories can provide a
'locked' approach to flood risk management, wherein the social norms of groups
with power presides over others and require significant effort and disturbance to
be altered.

Meanwhile, chapter 15 demonstrates how opinions of what *is* and what *ought* to
be at risk of flooding are differentially approached by different knowledge groups
in the USA. It explains how normativity can be shown to be positivity, embedding
normative information into insurance markets while excluding diverse and valid
knowledge sets with different normative beliefs and positive experiences. Such
action serves to decrease the efficacy of FRM approaches (in this example the
National Flood Insurance Program) and enhances social exclusion of those who are
not involved in knowledge generation.

Callon (1999) and Landstrom et al. (2011) demonstrate that an important step in
generating information that is both valid and trusted by epistemic communities is
to improve stakeholder engagement and overturn the paradigm of top-down
knowledge transfer from 'expert' to public. Stengers (2005) argues that more
dynamic knowledge exchange is required particularly in situations that are socio-
politically contentious – e.g. relating to flood risk, its sources and its reduction.

However, stakeholder engagement is a challenge, and not universally appropriate
or successful (see chapter 15; and Evers (2012)). The balance therefore comes in
matching information and knowledge flow systems with the political expectation
of the citizenry to be engaged with and understand FRM decisions. The pressure
and necessity for the two-way information exchange varies from country to
country as different ethoses of public engagement and information sharing exist.
Chapter 16 demonstrates the efficacy with which information spreads between
public communities during flood events via social media and how platforms such as
Twitter are increasingly used as a bridge between official and public communities.
The examples show how social media platforms as a component of FRM are
embraced differently in China and the USA, reflecting the approaches these states
take to the control and dissemination of information and the emphasis placed upon
the infallibility of state FRM.

Trust and accountability

Trust and accountability sit in a triangle with risk (Buchecker et al., 2013). Trust
and accountability are dependent upon perceptions of risk, and historical-cultural
experiences of flood events both individually and within society (Viklund, 2003).
Risk, trust and accountability are therefore historically and scientifically contingent,
sitting between individual behaviours and societal experiences.

Risk

The psychology of risk works to undermine constructivist arguments of human behaviour in relation to floods. A major critique of this social construction is that it fails to acknowledge and account for the biological reactions we have to certain situations (Stam, 2001). Indeed, Slovic et al. (2004) explain that in psychology and neuroscience we are considered to have two types of risk-based decision-making systems. Our *analytic system* is dependent upon algorithms, normative rules and calculus and requires conscious control to be enacted in a slow and effortful manner. It is, therefore, learned and socially constructed. Meanwhile our *experiential system* uses our automatic nervous system to make intuitive, quick and subconscious reactions. The latter system relies on emotional and evolutional associations with events and situations to make decisions. Chapter 14 expands on these explanations theoretically.

These systems are interdependent, as each system builds upon the other, and may sometimes clash in the decisions they prompt us to make. The combination of both analytic and experiential systems can help form a rational decision-making basis. On this psycho-neurological foundation, then, it is necessary for risk managers to build upon or mould individuals' experiential systems so that they accept and use risk information to inform their flood risk responses. Influencing the rational decision-making function requires trust in risk information, and encourages accountability from individuals to act to mitigate their flood risk when they are provided with appropriate information.

Trust

An important component of rational decision-making in relation to floods is the extent to which individuals and FRM organisations trust one another, not least because floods are not an everyday experience to be routinely observed (as one might observe teaching in a primary school) but come infrequently and perhaps unexpectedly. Institutional preparation and reaction during past disasters affects whether individuals feel that they can trust institutions for assistance before, during and after a particular flood event or whether they should take independent action to mitigate their flood vulnerability (Luo et al., 2015).

There is a paradox here in that effective institutional response to a flood hazard may engender trust but enhance societal risk by reducing the propensity of people to reduce their risk at an individual or household level basis (creating a moral hazard). Terpstra and Gutteling (2008) demonstrate this paradox in the case of households' perceived responsibilities for FRM in the Netherlands. However, lacking trust in institutions yet being unable to affect your own risk exposure is unfortunately a reality for many low-income households across the world. Indeed, it is often a privilege to be able to assess and react to your perceived flood risk when institutional trust might be low.

As discussed in this chapter's section on information and knowledge, public engagement techniques are often used to enhance flood risk communication mechanisms between 'experts' and 'lay' people. One significant outcome of such activities is that trust in risk-analysis and decision-making processes, and therefore their communicating bodies, can be enhanced. By involving publics in the process of computer flood risk modelling or in spatial planning decisions, engagement programmes seek to engender trust in institutional expertise and build consensus. However, turning to stakeholder engagement is not a panacea for building relationships of trust. Primarily, where programmes perform box-ticking exercises and outputs are not produced under consensus, trust may actually be eroded rather than developed, as participants become alienated and question project leaders' ability to listen (Buchecker et al., 2013). Engagement in FRM planning may enhance households' trust that decisions will be made by risk management organisations which are socially just and do not consider economic factors alone.

Accountability

A fundamental element of FRM is the diversion away from our historical reliance on nationally funded hard-engineering structures to reduce our flood risk (Tunstall et al., 2004). Such a dramatic paradigm shift has required social relations to floods to be reshuffled and renegotiated. When communities were predominantly reliant upon governments and environmental agencies to construct and maintain those flood defences, there was a clear direction in which to point fingers for who was accountable for creating and maintaining lower levels of flood risk. With broadly based FRM, and new combinations of engineering and non-structural measures such as spatial planning and matching flood insurance, the management of flood risk is spread through society, as Johnson and Priest (2008) and Butler and Pidgeon (2011) discuss in the UK context. Spreading accountability through society adds complexity both procedurally and legally and presents the requirement for expert institutions to develop frameworks that improve societal risk perception so as to relieve pressure on governments to protect its citizens through top-down and perhaps unaffordable engineered methods.

Trust, risk, and accountability exemplified

The status quo of situations most of the chapters in this volume describe is one where citizens and FRM stakeholders assume trust in government to make the right decisions for them as individuals and for them within the wider community. However, several chapters examine the implications when the relationship of trust is pressured or broken. Chapter 13 explores some of the issues when citizen–government trust is tested through emergency intentional flooding, particularly in situations where political discontent is exacerbated by poor FRM. Here, vulnerable communities with high reliance upon government decision-making for flood protection had their trust in government decision-making for 'protection' undermined. The case studies demonstrate, first, that trust constitutes an important and fragile part of historical relations between the Canadian government and indigenous communities, which is easily and damagingly broken; and, second, that the repercussions of losing the trust of vulnerable individuals in Thailand through poor FRM decision-making can be politically severe. This chapter prompts questions about the interrelation between fairness, trust and social justice.

Chapters 10 and 11 outline and exemplify the locations in which trust (via effective communication) is required to ensure functioning flood warning communication chains. Trust is required here to ensure that risk information is dispersed to the correct organisations at the correct time, and in a manner that is comprehensible to the recipients. Chapter 11 demonstrates the issues of accountability in Mugabe-era Zimbabwe, which experienced weak FRM governance and management procedures, forcing the finger of accountability to point towards institutional mismanagement when flood warnings failed. From a German perspective, chapter 2 examines the political and legal discussions of individual and governmental accountability for risk management through spatial planning and insurance coverage. Meanwhile, chapter 12 grapples with the issue of mixed accountability when home modifications are disallowed by residents in council-owned properties in the UK. Social constructions of property, individual choice and social good are at work to create a complex, contentious and apparently contradictory hydro-political landscape in which to carry out FRM.

Challenges of applying frameworks of reality and social constructionism to FRM

Challenges from inherent complexities

The field is a complex one, and social realities and constructions reflect this in political, economic, scientific, and environmental domains. In reality, as is well known (Smith, 2013), flooding comes from several sources: fluvial, coastal and 'surface' water flooding in urban conurbations and from farmland. Interventions to reduce risk can be quite different here: 'one-size fits all' approaches are at best ineffective and at worst dangerous. Particular complications arise when sources of flooding combine such that attributions are indistinct. Importantly, moreover, floods are episodic phenomena. Generally, the larger the flood the rarer it is and vice versa (Smith, 2013), and experiencing periods of no or minor floods can cause amnesia: flooding and its realities tend to occur at random through time and are therefore both unpredictable and forgettable.

The challenge from flooding is both that damaging and disrupting events occur now on a regular basis throughout the world – as do its social consequences – and also that risk appears to be increasing (Committee for Climate Change, 2016; Lazenby and Todd (2018) Penning-Rowsell et al., 2019), almost certainly as an anthropogenic effect (Kirtman et al., 2013). Sea level rise associated with glacier loss and changes in sea temperatures will also increase flood risk for low-lying communities (Hallegatte et al., 2013) (chapter 7) and there are particular challenges where rapidly urbanising areas are driving up risk levels (Oppenheimer et al., 2014) (chapter 13). Yet considerable uncertainty remains as to flooding's extent and spatial distribution (Beven and Hall, 2014).

As if understanding and assessing flood risk is not a complex enough matter, interventions have further complications (Parker, 2000) (Figure 1.1). Each can have several effects and impacts – economic, social and environmental – intended and unintended, singly or in combination. As debates concerning effective intervention measures have evolved, emphasis has grown on vulnerability reduction measures and away from engineering-only options (chapter 9). A recent trend also sees portfolios of measures that combine in creative ways to reduce risk, comprehensively tackling flood probability and exposure and vulnerability (Evans et al., 2004). We argue therefore, that strict attention must be paid to complexity when drawing socio-political conclusions for FRM planning.

Challenges to our ideas and our ontologies

Taking a hard constructionist approach (i.e. ignoring the plausibility of rationalism – see chapter 14) creates a possibility that we lose traction: we begin to dismiss phenomena as unreal or insignificant because they are a product of their social context or are socially unstable in their manifestation. However, constructionism encourages us to think about the ways in which a situation has arisen (chapter 5) and how social constructions have been moulded and to consider the knowledge that different experiences of an event can tell us about the way in which it could be better managed in the future.

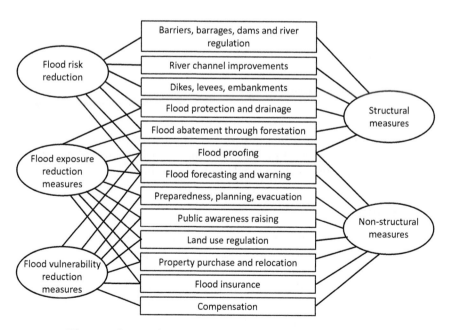

FIGURE 1.1 The complexity of interventions to reduce flood risk
Source: Parker, 2000.

Dismissiveness due to social constructionism makes assigning accountability harder as nothing is 'real'. For example, 'acceptable risk' is clearly a social construct – it is not fixed – but relational and probably different for both managers and flood victims. So, assigning accountability to either party becomes problematic because the very concept of acceptable risk is different for each. Both may have unrealistic expectations. Thus, social constructionism sees a situation where different people see things differently and have different views about the causes of things (such as risk) – and are therefore unlikely to accept responsibility or blame based on the other peoples' perceptions. This leads flood managers to cling to what they see as reality, not social constructions of reality, and thus absolve themselves of accountability. This is particularly pertinent to members of society who are not heard, understood or agreed with because they have different ideas of what is real to them (chapter 12).

There remains the question as to whether social constructionism and realism make us rely too much on normative/positivist accounts of the world. Does this duality ensure that a dichotomised world exists? Through setting subjective 'right' and a 'wrong' interpretations of flood events, normative thinking could undermine or ignore the opinions of those that experience floods differently. The 'other', i.e. the people who do not share the same experience as the societal group with the dominant flood narrative, may be ignored, discredited or struggle to have their voice heard. This clearly raises social justice issues. However, being attentive to this issue of power and environmental social justice can enable us to use social constructionism to establish where justice issues might emerge based on power imbalances (chapters 13 and 14).

Lastly, social constructionism as a framework of thought ignores the biological or physical attributes and constraints of the world that are out of our control (e.g. the seasons, tides, precipitation, topography, geology). It therefore abstracts power from the non-human to affect the human world. This is false and lends us to ignore the influence of the material environment on our societies. Our social relations are contingent upon material realities (e.g. the capacity of river basins, the porosity of local soils) and how we choose to live within them. Social constructionism removes some of the agency of the 'natural' world, and in doing so marginalises the attention that we should be paying to it.

Assessment

The discussion in this chapter exemplifies that it is not easy to define what is reality and what is constructed when you examine FRM, particularly when matters become techno-political and 'real' understanding is difficult for anyone but the specialist. Oftentimes, FRM is a messy entanglement of nature, science and society. Situations are dynamic: there are a number of challenges that will create pressure on relationships between different agents, shifting the field of trust and accountability with each flood event. Changes in power dynamics demonstrate a response to such challenges, perhaps with an emphasis on maintaining institutional stability

(Harries and Penning-Rowsell, 2010) while achieving sufficient flood mitigation levels, but often such changes are hidden from view (with decisions taken 'behind closed doors'). Questions of power to make such changes lead us to thoughts as to what are the right and wrong actions to take in given situations: what is fair, efficient and effective.

In this respect our analyses of floods allow us to examine and critique our own social norms, our assumptions of truths, and our relations to risk. Extremes here act as a 'magnifying glass' to illuminate – more than is normal – the relationship between humans and their environments.

The fifteen chapters that follow are grouped into three sections: governance issues; policy and implementation issues; and communities and people issues. This is a pragmatic arrangement, indicating one of the dominant themes within each chapter. But this categorisation is an imperfect over-simplification.

In reality, almost every chapter deals with some key governance issues – for example, the role of the state, regulations or markets – and each chapter has implications for the policies being pursued in the countries concerned, those policies setting the direction of travel of FRM there. Every example of flooding and FRM directly or indirectly concerns the communities and people affected: the impacts from the floods that they may have suffered, the measures that are designed to reduce the risk that they face, their attitudes to climate change, and the role of communities and individuals in implementing these measures and the policies that drive them. Each chapter also touches on aspects of FRM that can be seen as real and aspects that are best understood in terms of social constructions. The challenge is to see and to understand these differences, with all their complexities and fascinations.

References

Ailon, G. (2006). What B would otherwise do: A critique of conceptualizations of 'power' in organizational theory. *Organization*, 13(6), 771–800.

Allen, J. (2009). Three spaces of power: Territory, networks, plus a topological twist in the tale of domination and authority. *Journal of Power*, 2(2), 197–212.

Bachrach, P. and Baratz, M. S. (1962). Two faces of power. *American Political Science Review*, 56, 947–952.

Bankhoff, G., Frerks, G. and Hilhorst, D. (2004). *Mapping Vulnerability: Disasters, Development and People*. Routledge, London.

Beven, K. (2016). Facets of uncertainty: Epistemic uncertainty, non-stationarity, likelihood, hypothesis testing, and communication. *Hydrological Sciences Journal*, 61(9), 1652–1665.

Beven, K. and Hall, J. (eds) (2014). *Applied Uncertainty Analysis for Flood Risk Management*. Imperial College Press, London.

Birkland, T. A. (2016). Policy process theory and natural hazards. *Oxford Research Encyclopaedia of Natural Hazard Science*. Oxford, Oxford University Press.

Buchecker, M., Salvini, G., Baldassarre, G. D., Semenzin, E., Maidl, E. and Marcomini, A. (2013). The role of risk perception in making flood risk management more effective. *Natural Hazards and Earth System Sciences*, 13(11), 3013–3030.

Butler, C. and Pidgeon, N. (2011). From 'flood defence' to 'flood risk management': Exploring governance, responsibility, and blame. *Environment and Planning C: Government and Policy*, 29(3), 533–547.

Callon, M. (1999). The role of lay people in the production and dissemination of scientific knowledge. *Science, Technology and Society*, 4(1), 81–94.

Cardona, O.D. (2004). The need for rethinking the concepts of vulnerability and risk from a holistic oerspective: A necessary review and criticism for effective risk management. In Bankoff, G., Frerks, G. and Hilhorst, D. (eds), *Mapping Vulnerability: Disasters, Development and People*. Routledge, London.

Committee on Climate Change (2016). UK Climate Change Risk Assessment 2017: Synthesis Report: Priorities for the next five years. Available at www.theccc.org.uk/wp -content/uploads/2016/07/UK-CCRA-2017-Synthesis-Report-Committee-on-Clima te-Change.pdf

Dahl, R. A. (1957). The concept of power. *Behavioral Science*, 2(3), 201–215.

Demeritt, D., Nobert, S., Cloke, H. L. and Pappenberger, F. (2013). The European Flood Alert System and the communication, perception, and use of ensemble predictions for operational flood risk management. *Hydrological Processes*, 27(1), 147–157.

Emerson, R. M. (1962). Power-dependence relations. *American Sociological Review*, 27, 31–41.

Evans, E. P., Ashley, R., Hall, J. W., Penning-Rowsell, E. C., Saul, A., Sayers, P. B., Thorne, C. R. and Watkinson, A. (2004). Foresight flood and coastal defence project: Scientific summary. Volume 1, Future risks and their drivers. Office of Science and Technology, London.

Evers, M. (2012) Participation in flood risk management: An introduction and recommendations for implementation. Centrum för klimat och säkerhet, Karlstads universitet, Rapport 2012:1.

Faulkner, H., Parker, D. J., Green, C. and Beven, K. (2007). Developing a translational discourse to communicate uncertainty in flood risk between science and the practitioner. *Ambio*, 36, 692–704.

Foucault, M. (1972). *The Archaeology of Knowledge*. Sheridan Smith, A., trans. Pantheon, New York.

Foucault, M. (1979). *Discipline and Punish: The Birth of the Prison*. Sheridan, A., trans. Vintage Books, New York.

Hallegatte, S., Green, C., Nicholls, R. J., Corfee-Morlot, J. (2013). Future flood losses in major coastal cities. *Nature Climate Change*, 3, 802–806.

Harries, T. and Penning-Rowsell, E. C. (2010). Victim pressure, institutional inertia and climate change adaptation: The case of flood risk. *Global Environmental Change*, 21(1), 188–197.

Jasanoff, S. (2004). *States of Knowledge: The Co-production of Science and the Social Order*. Routledge, London.

Johnson, C. L., Tunstall, S.M., Penning-Rowsell, E. C. (2005). Floods as catalysts for policy change: Historical lessons from England and Wales. *International Journal of Water Resources Development*, 21(4), 561–575.

Johnson, C. L. and Priest, S. J. (2008). Flood risk management in England: A changing landscape of risk responsibility? *International Journal of Water Resources Development*, 24(4), 513–525.

Kirtman, B., Power, S. B., Adedoyin, J. A., Boer, G. J., Bojariu, R., Camilloni, I., Doblas-Reyes, F. J., Fiore, A. M., Kimoto, M., Meehl, G. A., Prather, M., Sarr, A., Schär, C., Sutton, R., van Oldenborgh, G. J., Vecchi, G. and Wang, H. J. (2013). Near-term climate change: Projections and predictability. In Stocker, T. F., Qin, D., Plattner, G.-K., Tignor, M., Allen, S. K., Boschung, J., Nauels, A., Xia, Y., Bex, V. and Midgley, P. M. (eds), *Climate Change 2013: The Physical Science Basis. Contribution of Working Group I to the Fifth Assessment Report of the Intergovernmental Panel on Climate Change*. Cambridge University Press, Cambridge, 953–1028.

Landström, C., Whatmore, S.J., Lane, S. N., Odoni, N. A., Ward, N. and Bradley, S. (2011). Coproducing flood risk knowledge: Redistributing expertise in critical 'participatory modelling'. *Environment and Planning A*, 43(7), 1617–1633.

Lazenby, M. J. and Todd, M. C. (2018). Future precipitation projections over central and southern Africa and the adjacent Indian Ocean: What causes the changes and the uncertainty? *Journal of Climate*, 31, 4807–4826.

Lukes, S. (2004). *Power: A Radical View*. 2nd edition. MacMillan, London.

Lupton, B. (1999). *Risk*. Taylor & Francis, Abingdon, UK.

Luo, T., Maddocks, A., Iceland, C., Ward, P. and Winsemius, H. (2015). World's 15 countries with the most people exposed to river floods. World Resources Institute. Available at www.wri.org/blog/2015/03/world%E2%80%99s-15-countries-most-people-exposed-river-floods

Oppenheimer, M., Campos, M., Warren, R., Birkmann, J., Luber, G., O'Neill, B. and Takahashi, K. (2014). Emergent risks and key vulnerabilities. In Field, C. B., Barros, V. R., Dokken, D. J., Mach, K. J., Mastrandrea, M. D., Bilir, T. E., Chatterjee, M., Ebi, K. L., Estrada, Y. O., Genova, R. C., Girma, B., Kissel, E. S., Levy, A. N., MacCracken, S., Mastrandrea, P. R. and White, L. L. (eds), *Climate Change 2014: Impacts, Adaptation, and Vulnerability. Part A: Global and Sectoral Aspects. Contribution of Working Group II to the Fifth Assessment Report of the Intergovernmental Panel on Climate Change*. Cambridge University Press, Cambridge, 1039–1099.

Parker, D.J. (ed.) (2000). *Floods*. Routledge, London.

Penning-Rowsell, E. C., Priest, S., Parker, D. J., Morris, J., Tunstall, S., Viavattene, C. and Owen, D. (2013). *Flood and Coastal Erosion Risk Management: A Manual for Economic Appraisal*. Routledge, London.

Penning-Rowsell, E. C. and Korndewal, M. (2018). The realities of managing uncertainties surrounding pluvial urban flood risk: An ex post analysis in three European cities. *Journal of Flood Risk Management*. Available via https://doi.org/10.1111/jfr3.12467.

Penning-Rowsell, E. C., Pardoe, J., Hall, J. and Self, J. (in press 2019). Policy processes in flood risk management: Three theories and four case studies. In Dadson, S. et al. (2019), *Water Science, Policy and Management*. Wiley-Blackwell, London,

Petts, J. (2008). Public engagement to build trust: False hopes? *Journal of Risk Research*, 11(6), 821–835.

Pfohl, S. (2008). The reality of social constructions. In Holstein, J. and Gubrium, J. (eds), *Handbook of Constructionist Research*. Guilford Publications, New York and London.

Rowe, G. and Frewer, L. J. (2000). Public participation methods: a framework for evaluation. *Science, Technology, & Human Values*, 25(1), 3–29.

Self, J. and Penning-Rowsell, E. C. (2017). Power and policy in floodplain management, drawing on research in Alberta, Canada. *Water Policy*, 20, 21–36.

Slovic, P., Finucane, M. L., Peters, E. and MacGregor, D. G. (2004). Risk as analysis and risk as feelings: Some thoughts about affect, reason, risk, and rationality. *Risk Analysis*, 24 (2), 311–322.

Smith, K. (2013). *Environmental Hazards: Assessing Risk and Reducing Disaster*. 6th edition. Routledge, London.

Stam, H. J. (2001). Introduction: Social constructionism and its critics. *Theory and Psychology*, 11(3), 291–296.

Stengers, I. (2005). The cosmopolitical proposal. In Latour, B. and Weibel, P. (eds), *Making Things Public: Atmospheres of Democracy*, 994.

Terpstra, T. and Gutteling, J. M. (2008). Households' perceived responsibilities in flood risk management in the Netherlands. *International Journal of Water Resources Development*, 24(4), 555–565.

Tunstall, S. M., Johnson, C. L. and Penning-Rowsell, E. C. (2004). Flood hazard management in England and Wales: From land drainage to flood risk management. *World Congress on Natural Disaster Mitigation*, 18–21 February 2004, New Delhi, India.

Viklund, M. J. (2003). Trust and risk perception in Western Europe: A cross-national study. *Risk Analysis*, 23(4), 727–738.

Weber, M. (1964). *The Theory of Social and Economic Organizations*. The Free Press, New York.

2

LEGAL GEOGRAPHY AND FLOOD RISK MANAGEMENT IN GERMANY

Matilda Becker

Introduction

Flood risk management (FRM) has overtaken the use of flood defences as the primary flood management approach in Western Europe since the mid-1990s (Johnson and Priest, 2008; Butler and Pidgeon, 2011; Krieger 2013). FRM focuses on integrated, cross-disciplinary organisation, where stakeholder views are assimilated into management plans at the local scale within river basin districts. With risk at its core, FRM relies upon the quantification of hazards, pre-emptive decision-making and appropriate recovery plans (Krieger, 2013).

Particular geographies bring additional challenges. As demonstrated by Cook (2015) in Canada and Garrick et al. (2014) in the USA and Australia, dealing with transboundary resources such as water requires inter-scalar and inter-institutional political and organisational collaboration. Such collaboration can also prove problematic within federal systems where states may experience unequal risks of flooding and have mixed political agendas and different human-ecological histories (Moss, 2004; Hartmann and Albrecht, 2014). The diversity of federal systems, both politically and socio-culturally, provides added complexities for creating consistent and integrated river basin management plans. This is demonstrated here by FRM implementation in federal Germany (Hartmann and Spit, 2016).

Several factors caused Germany with its 16 federal states (singular Land; plural Länder) to transition from a flood defence to a FRM approach. First, severe flooding in the 1990s and 2000s (e.g. Rhine 1995, Oder 1997, Elbe 2002) undermined the assumed efficacy of flood defences to protect homes and infrastructure from damage as well as emphasising the increased technical and financial costs of construction and maintenance (Krieger, 2013). Second, European Union (EU) directives – including the Water Framework Directive (WFD) in 2000 and Floods Directive (FD) in 2007[1] – legally required shifts towards the FRM paradigm

(EC, 2000/60/EC; EC, 2007/60/EC). Passages from the Floods Directive were incorporated directly into the 2009 German National Water Law (WHG). Länder thereby became legally responsible for implementing FRM as promulgated by the EU directive (WHG, 2009). Third, FRM is also a reaction to global climate change, which threatens to increase the extent and severity of flood events in Germany as elsewhere, reducing the ability of existing flood defences to protect land, people, and property (Hattermann et al., 2014; Hartmann and Albrecht, 2014; Madsen et al., 2014).

The shift towards FRM and the implementation of a diversity of risk management techniques have not been uniform throughout the country. With water governance controlled at Länder level, Germany is a mosaic of different management structures and mechanisms. Until reunification in 1990, East Germany's rivers were governed on the basin scale, while West Germany governed its rivers at Länder level. After reunification, East Germany's spatial governance structure was aligned with that of West Germany. This lasted until 2000, when the WFD required the reorganisation of management into River Basin Districts.[2] The transition required significant institutional adaptation to meet the new EU requirements (Moss, 2004).

For FRM, and implementation of the FD, the WFD was critical in setting geographical precedents for organisation. There are clear overlaps between the maintenance of good water quality, ecosystem services provision, and the protection from flood risk. There is also significant space for conflict (e.g. flooding serves to provide a diversity of hydro-ecologies, yet threatens water quality provision and public safety). Therefore, an integration of River Basin Management Plans (WFD) and Flood Risk Management Plans (FD) is both logical and necessary for effective water management that adheres to legal requirements. The implementation of the FD is hugely reliant upon the spatial-legal structures promulgated in 2000 by the WFD, which provided the institutional ordering of management responsibilities within river basin districts.

Applying legal geographies to this German situation enables a focus on the legal approach to managing risk in space, and begins to untangle a messy story of mixed responsibilities between government, private companies, and citizens. This chapter highlights issues with the transition to localised responsibilities, and tensions between Länder and Federal governments for environmental, political, and financial decision-making power.

Legal geographies as a framework of analysis

Legal geography 'investigates the co-constitutive relationship of people, place, and law' (Bennett and Layard 2015, 406). Law is reliant upon geography and space for its creation, as much as space is reliant upon the law for its socio-political ordering (Braverman et al., 2014). For this chapter, the geography of flooding will primarily refer to geographies of risk.

Law has a *de facto* and *de jure* impact on how people interact with and order space. Key literature has highlighted the political-ecological relation between people and space, with an emphasis on property (see for example Delaney, 1998; Blomley 2003; Blomley 2007; Kedar et al. 2018). Legal geographies are often used in the analysis of the day-to-day spatial manifestations of the law, for example in relations between neighbours' gardens (Blomley 2014) and the regulation of trees in cities (Braverman 2008).

Legal geography is a 'rich yet eclectic' subject (Braverman et al. 2014, 2). However, given that it is prescribed as an investigation of space and law, a large degree of flexibility is offered in its manner of application. We have seen no application of legal geographies to the context of flood risk management, where an examination of clashing or complementing legal frameworks – that can hinder or support FRM measures – should be fruitful and identify the challenges and successes in implementing FRM plans effectively.

Legal geographies can be successfully applied to the German context where geography (or better, the river basin) has been defined through law as the appropriate space through which to govern and control water for flood management. Political boundaries do not suffice in a water management model, which must account for the integrated, multiple and co-dependent hydro-ecological infrastructures within basins. Therefore, international European law has necessitated for hydrological management to be executed at the river basin level (EC, 2000/EC/ 60), whether or not the basin crosses internal or international political boundaries.

Conversely, Butler and Pidgeon (2011), Doorn (2016), and Oels (2006) have argued that FRM has caused the down-scaling of management responsibility to be shifted to the site of local governmental institutions, NGOs, private organisations and individuals (see also Penning-Rowsell and Johnson (2015)). Doorn (2016) uses the Rawlsian analysis of distributional justice to review international FRM policy on how responsibility for FRM has been dispersed through societies, noting that typically in developed countries, responsibility for flood risk management and resilience is judicially placed upon citizens, with government and emergency services stepping into action during and after a flood event or in major protection schemes. Therefore, the way river basin management plays out in practice, both spatially and temporally, and how the law wrestles between state and individual responsibilities for FRM can be examined in this German example.

Oels (2006), Butler and Pidgeon (2011), Rose (1996) and Penning-Rowsell and Johnson (2015) argue that Western governance has become increasingly decentralised, with greater involvement of local institutions and organisations as well as citizens and communities. While power is still centrally held, responsibility and accountability for management and risk is decentralised to the local level. Public participation schemes will enable individuals to 'fulfil themselves within a variety of micro-moral domains or communities' (Rose, 1996, 57). Furthermore, individuals are encouraged to become autonomous and self-optimising, risk averse and calculative in their decision-making (see also chapter 12).

As stated in the German 2009 WHG (National Water Law), the public are expected to do whatever is 'possible and reasonable' to reduce flood risk to themselves and their property (WHG, 2009, §5 (2)). However, in Germany, devolution of accountability to citizens has not occurred evenly. This is demonstrated here with three examples: spatial planning; private flood insurance; and public participation in FRM. Spatial planning and private insurance have been chosen because of their importance to FRM and common use as effective and sustainable ways of managing flood risk in developed countries. Public participation has been chosen because of the growing academic and governmental interest in engaging stakeholders in environmental management.

An interesting friction is at work between flood management plans working to reduce risk at the local and regional level, and laws to do with spatial planning, home insurance and risk communication. These are instances, analysed below, where soft/non-structural flood risk management measures have been (attempted to be) implemented in Germany to reduce community and individual flood risk but clashed with constitutional rights of the individual, particularly in relation to free choice and property.

Spatial planning within flood risk management

The intersection of spatial planning with flood risk management demonstrates best how the law responds to space and scale. Tarlock and Albrecht (2018) identify this intersection as one area where laws to reduce flood vulnerability clash with constitutional laws and private property rights. Property rights are often deemed a fundamental right (Waldron, 2012), on the Lockean principle that labour exerted on land makes it a person's property. However, Tarlock and Albrecht (2018) explain how, despite traditional theories of individual entitlement, property rights entail social obligations (i.e. something which does not inhibit the public good[3]) under the German constitution. Therefore, German courts consistently uphold the principle that private land may be interfered with by government if it is in the public interest, where continuation of private ownership would impinge on its component of the public good.

This principle responds to the fact that the flood risk of inhabitants and users of the river basin is contingent upon changes within the river basin. The WHG states that Länder are forbidden to carry out land-use/river alterations which place other regions within their river basin at risk (WHG, 2009, §75 (4)). The government can take land out of private hands (not always with compensation) to reduce/mitigate the flood risk of others (Tarlock and Albrecht, 2018). A famous example setting the precedent for this comes from Hamburg in 1968, where all grassland in the city-state was converted from a private to a public good to work as flood plains (BVerfG, Ruling of 18.12.1968). The Hamburg case demonstrates an instance where the full rights to property are revoked. However, it need not always be so absolute, with restrictions on property owners' use of their land often sufficing for the purposes of FRM. This might involve, for example, the restriction of construction rights on their land (BVerwG, Ruling of 22.07.2004). This has been reinforced again by §78 of the WHG, which forbids building on flood plains in nine different instances but allows for exceptions when the wellbeing of the citizenry is not at risk, nor the risk from flooding elevated by such development (e.g. by using flood-safe architecture).

The federal government mandates how and where Länder can build in relation to flood plains. Thus, the legal power is centralised to the federal government, yet the Länder retain responsibility for implementing land use planning at the regional level. An example of the subtlety of planning that is allowed for is demonstrated in the WHG (§76d (2)), which states that Länder can set their own criteria for zones that are thought to develop flash floods based on topographical and hydrographical information. Here is a clear example of where spatial planning for FRM must respond to the risk geographies of an area and where the law explicitly acknowledges the importance of hydro-geography in policy-making.

Spatial planning is a legal intervention into the routine of flood plain management and FRM planning. However, the following example of home insurance demonstrates how the intervention into the day-to-day of individual citizens' flood protection is not deemed a sufficient social good to warrant promulgation as law and the surpassing of individual rights.

Home insurance as flood risk management

Law and geographies of risk can be well examined through insurance. Here, geography directly affects the ability of homeowners to address their financial vulnerability by taking out risk-based insurance cover. Insurance is therefore a key means of accepting self-responsibility for flood risk. Flood insurance is not compulsory in Germany and is not included in extension packages for natural disasters in home insurance policies; it must therefore be purchased as a separate add-on. Attempts to introduce compulsory flood insurance as federal legislation have failed.

From compulsory to optional flood insurance

Coverage of flood insurance is poor and uneven in Germany, with only 41 per cent of households being policy owners (GDV, 2018). This could be explained by 93 per cent of the sample population falsely believing that they are automatically covered for flood events (GDV, 2016). The new federal Länder (ex-East German states)[4] and Baden-Württemberg (95%) in the west have the highest coverage rates (Table 2.1). Historically, flood damage was included automatically under household insurance in East Germany (for which there was 95% penetration in 1978) (Bader, 1980). Remaining policies were transferred to private insurers post-reunification (Thieken et al., 2006), which explains some of the residual high penetration levels in the new federal Länder.

Similarly, in Baden Württemberg flood insurance came with compulsory state-provided home insurance, but was abandoned in 1994 under EU anti-monopolisation laws for state-owned housing insurance products (Lange, 2011). In 1996, Epple and Schäfer (1996) argued that the dissolution of over 200 years of state-provided home-insurance would have significant impacts on the cost and therefore

TABLE 2.1 Insurance coverage in Germany by Land. Länder in bold were previously part of the German Democratic Republic (East Germany). Data adapted from GDV, 2018.

Land	Insurance coverage (%)
Baden-Württemberg	94
Saxony	**47**
Thuringia	**46**
Saxony-Anhalt	**43**
North Rhine-Westphalia	42
Hesse	36
Berlin	**33**
Brandenburg	**33**
Bavaria	32
Rhineland-Palatinate	31
Mecklenburg Western-Pomerania	**25**
Saarland	25
Hamburg	21
Lower Saxony	20
Bremen	19

penetration of natural disaster insurance in homes. However, despite their warnings, penetration remains at 94 per cent. While the Baden-Württemberg example clearly demonstrates the interference that EU competition laws have in state-run financial institutions, one must question why Baden-Württemberg's penetration remains so high, while former East German states' coverage more than halved in the same period.

It is possible that because the original state insurer did not charge premiums proportional to homeowners' flood risk, but instead required 'contributions' equal throughout all homeowners, incoming private insurers may have struggled to provide coverage while offering unattractive premiums to a market which was historically used to low insurance costs. In relation to the lower penetration in the former East German states, the disastrous impacts of the 2002 Elbe floods almost doubled the number of uninsurable properties, thus decreasing the proportion of homeowners purchasing flood insurance (Schwarze and Wagner, 2007).

Failure to re-introduce nationwide compulsory flood insurance

One mechanism used to ensure the solvency of the flood insurance industry is through requiring universal coverage (e.g. as provided through public-private insurance in France). This ensures that coverage is spread through the population, and makes insurance companies more resilient to claims after a flood. Chapter 3 discusses the criteria necessary for viable flood insurance.

A difficulty encountered in attempting to universalise flood insurance in Germany was a clash between the rights of the individual under the German constitution and the attempt to enforce compulsory flood insurance upon homeowners. The attempt to pass this bill was a result of the 2002 Elbe floods, which caused massive insurance claims. Since a high proportion of all policies held in Germany were in the Elbe basin, a disproportionate impact was felt by insurers and the national government (Schwarze and Wagner 2004, 2007). With damages totalling €11.6 billion, compensation from insurers was bolstered by significant government financial aid to compensate the uninsured (€7.6 billion) (Thieken et al., 2006). The bill was not passed by the Bundestag (the lower house of the German parliament).

A primary condition under which the constitutional right to individual autonomy can be broken is where a decision is made in the public interest (see the public/social good point above). This, one would think, is similar to the social obligation of property, described in the previous section. Indeed, Schwarze and Wagner (2007) propose that a national flood insurance programme would be in the public interest, given that an imbalance in insurance policies nationally leaves the German economy vulnerable to needing to deliver aid to those who would otherwise receive payments from their insurers. Aid payments would, ultimately, be provided by the taxpayer and restrict spending in other public sectors.

Schwarze and Wagner (2007) argue that the inability to frame the legality of the proposal for universal insurance is short-sighted and lacks the foresight to create financial resilience under the growing shadow of climate change and increasingly severe flood events. Therefore, in this instance, financial resilience was not deemed a sufficient public good to warrant the interference into individual choice by government. Another sticking point in the universalisation of flood insurance is the geographical irregularity of flooding throughout Germany, which meant those states with low flood risk rejected the proposal (as they would have to contribute financially), which did not benefit them directly. A pattern of opposition is similar to the pattern of insurance omission in Table 2.1. However, chapter 15 in this volume demonstrates that compulsory flood insurance is not necessarily a panacea, using the United States as an example.

The lack of support for compulsory flood insurance perhaps indicates an element of proportionality and relativity in the assignment of 'public good' to FRM measures. Private property can be made public property or have development restrictions placed upon it in areas where a demonstrable impact on public safety from floods is present. It is therefore a geographically selective and risk-proportional interference into individuals' rights. Universal flood insurance is however absolute and non-relative to actual risk exposure in its coverage.

Therefore, the issue of private home insurance brings into question how well law can govern space and is attuned to the growing risks attributed to climate change. It highlights the competing legalities with which flood risk management must work, and the balance of financial, constitutional, and federal responsibilities that are weighed up in the definition of spatial management laws. Due to the incompatibility of compulsory flood insurance with German legal norms, the country remains over-reliant on other soft/non-structural management techniques. Homeowners are relied upon to make changes to their properties and purchase insurance, encouraged via communication methods of flood risk mapping tools and public participation projects.

Public participation in flood risk management

Examples of theorising on flood risk management demonstrate that an engagement of diverse publics in environmental management leads to the expansion of knowledge within communities, and indeed appreciation by the 'lay' publics of the flood risk at hand (Callon 1999; Stengers, 2005; Landström et al., 2011). The engagement of a diversity of knowledges and experiences ensures that the geography of risk is (hopefully) better understood and reacted to holistically (Evers, 2012). See chapter 15 for greater detail about the benefits and drawbacks of engaging 'the public' in flood risk management planning.

Important aspects of 'public' engagement include *who* the public is, *how* they should be engaged with, and at what *stage* in the environmental decision-making process this should occur. Public engagement/participation for FRM is vague at the European legal scale in terms of addressing these parameters (Albrecht, 2016). The Floods Directive outlines the need to engage 'publics' in Articles 9 and 10, and this has been translated directly into the German WHG (2009). While the WFD stipulates that the public must be able to respond to plans for at least six months, the FD does not provide such a timescale. Furthermore, the FD does not define where or how publics should be informed (i.e. via what fora). Such definition is left for member states to specify in their legal translations of the Directive.

In Germany, public participation – or the engagement of the 'concerned public' – as defined by the Federal Environmental Protection Act (UVPG, 2010) is a legally vague requirement with huge leeway for development. Newig et al. (2014) explain how this legal vagueness is the response to opposition by Länder to stricter public engagement laws. This vagueness is an overriding feature of the WHG §79, and as such relies upon state legislation to define in more specific terms their public engagement, and on the UVPG and the Federal Administrative Procedures Act (VwVfG, 2003) to create clarity on what is required. First, the inclusion of only 'concerned' publics is meant to ensure that only those stakeholders who would be directly affected by flood risk management plans should have a say on their promulgation. This is beneficial in that it prevents policy opponents from using the flood plans as a proxy to contest power (Albrecht, 2016), but there is the severe drawback of 'directly affected' still being a vague concept – does this mean affected physically, economically, socially or culturally?

Furthermore, within the WHG, no time period for public engagement is outlined for flood risk management. However, the UVPG necessitates that at least one month must be given for concerned publics to respond to flood risk management plans. Länder adhere to this requirement but can extend the consultation period beyond one month. However, any review of public engagement projects demonstrates that they are inherently time-intensive, with one month providing limited time for the dissemination and collection of materials to and from the public in an effective manner that does not just simply constitute a box-ticking exercise. In addition to the *period* of consultation, the *stage* in the flood risk management process (see note 1) during which the public must be consulted is restricted to the post-flood risk mapping stage (stage 4).

Therefore, there is no legal requirement for member states to engage their publics in the creation of flood risk maps. Landström et al. (2011) and Becker et al. (2017) give an outline of how public involvement in risk mapping can enhance both specialist and public understanding of risk and how it may be mitigated. This example of temporality in law in Germany expands on the focus of law and its day-to-day manifestation in people's lives (e.g. the literature cited earlier in the 'Legal geographies' section of this chapter), and expands it to encompass a discussion of how people are organised temporally within long-term project planning, situating publics within wider administrative/organisational scales and offering space for new discussions of power and influence.

Lastly, in terms of *where* and *how* flood risk plans must be posted for response, there is no stipulation within the WHG, nor in the UVPG, in terms of location. However, the Administrative Procedures Act (2003) places the requirement upon Länder to post all documents and consultation timetables online in a public location (i.e. on the Länder's Environment website, or similar).

The three aspects of public engagement outlined at the beginning of this section demonstrate that Germany has developed a bare-minimum approach to public engagement and left significant room for Länder to promulgate their own engagement rules. The engagement framework employed in Germany is at best weak. However, as Wynne (1996) and Callon (1999) describe, states prefer to direct attention to address risk through expert/formal channels of measurement and management. Germany demonstrates that the individual – or the mass which 'the public' represent – are not deemed necessary contributors within the risk management planning framework.

Conclusions

Legal geographies in this German case study enables a focus on the legal approach to managing risk in space and begins to untangle a messy story of mixed responsibilities between government, private companies and citizens. Far from representing a unified and consistent approach to risk management, Germany demonstrates a pattern fractured along political-geographic and historical-economic lines. Asking *why* legal approaches to FRM in Germany have thus far failed to create consistent (and perhaps socially just) patterns of risk management reveals clashes between legal rulings at multiple jurisdictional scales and a pattern of vague legal statements relating to responsibilities to disseminate information to the general public.

Furthermore, legal geographies applied to FRM in Germany highlight two key issues where the individual is at the centre of both. First, there is an irregularity, both spatial and legal, in the definition of individuals' rights and the instances where the government may override them for the public good. Second, there is a lack of clarity, and therefore discord and diversity nationally, as to the extent to which the public should be engaged in FRM planning.

These examples raise questions about what the importance of the individual is in FRM and where, when and if the right of the individual should be overridden for the public interest. There are clear social justice issues when utilitarian models begin to be contemplated. However, global climate change is exacerbating the risk to citizens of floods, with impacts distributed unevenly through society both spatially and ethnographically. Additionally, the mechanisms via which the implications of floods affect citizens nationally (e.g. through increased government expenditure on national aid) means that ultimately, protection from the implications of flood risk should be considered a social obligation for government and its agents, rather than a right of an individual to opt out.

Public engagement demonstrates a means via which citizens may be empowered to understand and act upon their flood risk, which is the antithesis to the infringement of individual rights. However, this German case study demonstrates that a two-pronged approach is required, first to engage citizens in thinking about their flood risk, its geography and how to act to mitigate it, and second, to ensure that the legal and financial infrastructures are in place at the appropriate spatial scale to make different flood vulnerability reduction measures viable for households (e.g. by reducing the cost of flood insurance to the individual through national-level universalisation).

Notes

1 The 2007 Floods Directive (EC, 2007/60/EC) requires member states to have submitted draft plans by end-2015 to:
    i   Carry out flood plain identification in their basin areas
    ii Carry out assessments of risk of potential flooding in their basin areas
    iii Create flood hazard and flood risk maps of their basin areas
    iv Create flood risk management plans based on the above and coordinate appropriately with other member states sharing the river basin.
    These criteria have been copied into the 2009 German national Water Law.
2 With the exception of North Rhine-Westphalia, which has managed its rivers on a basin level since before the WFD in 2000.
3 Here, public good follows the non-economic, non-legal definition of a service that provides benefits at a societal level (i.e. this definition is *not* after Ostrom, 1990). Sometimes referred to as 'social good'.
4 Ex-East German states Saxony (47%), Thuringia (44%), and Saxony-Anhalt (43%)

References

Albrecht, J. (2016). Legal framework and criteria for effectively coordinating public participation under the Floods Directive and Water Framework Directive: European requirements and German transposition. *Environmental Science and Policy*, 55, 368–375.

Bader, H. (1980). *Die staatliche Versicherung in der DDR: Sach-, Haftpflicht- und Personenversicherung.* Die Wirtschaft Verlag, Berlin.

Becker, M., Odoni, N., Landström, C. and WhatmoreS. J. (2017) Community Modelling: Incorporating local knowledge into hydraulic flood modelling in Otley, Yorkshire (downloadable from www.communitymodelling.org/projects-2-the-process.html).

Bennett, L. and Layard, A. (2015). Legal geography: Becoming spatial detectives. *Geography Compass*, 9(7), 406–422. Available at http://doi.wiley.com/10.1111/gec3.12209.

Blomley, N. (2003). Law, property, and the geography of violence: The frontier, the survey, and the grid. *Annals of the Association of American Geographers*, 93(1), 121–141.

Blomley, N. (2007). Making private property: Enclosure, common right and the work of hedges. *Rural History*, 18(1), 1–21.

Blomley, N. (2014). Learning from Larry: Pragmatism and the habits of legal space. In: Braverman, I., Blomley, N., Delaney, D. and Kedar, A. (eds), *The expanding spaces of law: a timely legal geography*. Stanford Law Books, Stanford, 77–94.

Braverman, I. (2008). Governing certain things: The regulation of street trees in four North American cities. *Tulane Environmental Law Journal*, 22, 1–26.

Braverman, I., Blomley, N., Delaney, D. and Kedar, A. (2014). Introduction: Expanding the spaces of law. In: Braverman, I., Blomley, N., Delaney, D. and Kedar, A. (eds), *The expanding spaces of law: A timely legal geography*. Stanford Law Books, Stanford, 1–29.

Butler, C. and Pidgeon, N. (2011). From 'flood defence' to 'flood risk management': Exploring governance, responsibility, and blame. *Environment and Planning C: Government and Policy*, 29, 533–547.

BVerfG, Ruling of 18.12.1968–1 BvR 638, 673/64 and 200, 238, 249/65, BVerfGE 24, 387 ff. 1968.

BVerwG, Ruling of 22.07.2004–7 CN 1.04. Bundesverwaltungsgericht. Available from www.bverwg.de/220704U7CN1.04.0 [Accessed 29. 06. 2018]

Callon, M. (1999). The role of lay people in the production and dissemination of scientific knowledge. *Science Technology and Society*, 4(1), 81–94.

Cook, C. (2015). Getting to multi-scalar: An historical review of water governance in Ontario, Canada. In Norman, E. S., Cook, C. and Cohen, A. (eds), *Negotiating water governance: Why the politics of scale matter.* Ashgate Publishing, Farnham.

Delaney, D. (1998). *Race, place and the law*. University of Texas Press, Austin.

Doorn, N. (2016). Distributing responsibilities for safety from flooding. *E3S Web of Conferences*, 7(24002), 1–6.

EC (European Community) (2000). Directive 2000/60/EC of the European Parliament and of the Council of 23 October 2000 establishing a framework for community action in the field of water policy. *Official Journal of the European Communities*, L327, 1–72.

EC (European Community) (2007). Directive 2007/60/EC of the European Parliament and of the Council of 23 October 2007 on the assessment and management of flood risks. *Official Journal of the European Communities*, L228, 27–34.

Epple, K. and Schäfer, R. (1996). The transition from monopoly to competition: The case of housing insurance in Baden-Württemberg. *European Economic Review*, 40(3–5), 1123–1131.

Evers, M. (2012). Participation in flood risk management: An introduction and recommendations for implementation. Rapportserie Klimat och säkerhet 2012:1, Karlstads Universitet, Karlstads.

Garrick, D., Anderson, G.R.M., Connell, D. and Pittock, J. (2014). *Federal rivers: Managing water in multi-layered political systems*. Edward Elgar, Cheltenham.

GDV (Gesamtverband der Deutschen Versicherungswirtschaft e. V.) [Association of German Insurers] (2016). Die wichtigsten Umfrageergebnisse zum Naturgefahrenschutz im

Überblickwww.gdv.de/de/themen/news/die-wichtigsten-umfrageergebnisse-zum-natur gefahrenschutz-im-ueberblick-12172. [Accessed 29. 06. 18]

GDV (Gesamtverband der Deutschen Versicherungswirtschaft e. V.) [Association of German Insurers] (2018). Mehrheit der Gebäude in Deutschland nicht richtig gegen Naturgefahren versichert. www.gdv.de/de/themen/news/mehrheit-der-gebaeude-in-deutschland-nicht-richtig-gegen-naturgefahren-versichert-12176. [Accessed 21. 06. 18]

Hartmann, T. and Albrecht, J. (2014). From flood protection to flood risk management: Condition-based and performance-based regulations in German water law. *Journal of Environmental Law*, 26(2), 243–268.

Hartmann, T. and Spit, T. (2016). Implementing the European flood risk management plan. *Journal of Environmental Planning and Management*, 59(2), 360–377.

Hattermann, F. F., HuangS., Burghoff, O., Willems, W., Österle, H., Büchner, M. and Kundzewicz, Z. (2014). Modelling flood damages under climate change conditions: A case study for Germany. *Natural Hazards and Earth System Sciences*, 14, 3151–3169.

Johnson, C. L. and PriestS. J. (2008). Flood risk management in England: A changing landscape of risk responsibility. *International Journal of Water Resources Development*, 24, 513–525.

Kedar, A., Amara, A. and Yiftachel, O. (2018). *Emptied lands: A legal geography of Bedouin rights in the Negev*. Stanford University Press, Stanford.

KriegerK. (2013). The limits and variety of risk-based governance: The case of flood management in Germany and England. *Regulation and Governance*. 7, 236–257.

Landström, C., Whatmore, S. J., Lane, S. N., Odoni, N. A., Ward, N. and Bradley, S. (2011). Co-producing flood risk knowledge: Redistributing expertise in critical 'participatory modelling'. *Environment and Planning A*, 43(7), 1617–1633.

Lange, T. (2011). Die (Pflicht-) Versicherung von Elementarrisiken in Deutschland. Universitätsverlag Göttingen, Göttingen.

Moss, T. (2004). The governance of land use in river basins: Prospects for overcoming problems of institutional interplay with the EU Water Framework Directive. *Land Use Policy*, 25, 85–94.

Madsen, H., Lawrence, D., Lang, M., Martinkova, M. and Kjeldsen, T. R. (2014). Review of trend analysis and climate change projections of extreme precipitation and floods in Europe. *Journal of Hydrology*, 519(D), 3634–3650.

Newig, J., Challies, E., Jager, N. and Kochskämper, E. (2014). What role for public participation in implementing the EU Floods Directive? A comparison with the Water Framework Directive, early evidence from Germany and a research agenda. *Environmental Policy and Governance*, 24(4), 275–288.

Oels, A. (2006). Rendering climate change governable: From biopower to advanced liberal government? *Journal of Environmental Policy and Planning*, 7(3), 185–207.

Ostrom, E. (1990). *Governing the commons: The evolution of institutions for collective action*. Cambridge University Press, Cambridge.

Penning-Rowsell, E. C. and Johnson, C. L. (2015) The ebb and flow of power: British flood risk management and the politics of scale. *Geoforum*, 1(62), 131–142.

Rose, N. (1996). Governing 'advanced' liberal democracies. In Barry, N., Osborne, T. and Rose, N. (eds), *Foucault and political reason: Liberalism, neo-liberalism and rationalities of government*. Routledge, London.

Schwarze, R. and Wagner, G. G. (2004). In the aftermath of Dresden: New directions in German flood insurance. *The Geneva Papers on Risk and Insurance*, 29(2), 154–168.

Schwarze, R. and Wagner, G. G. (2007). The political economy of natural disaster insurance: Lessons from the failure of a proposed compulsory insurance scheme in Germany. *European Environment*, 17, 403–415.

Stengers, I. (2005). The cosmopolitical proposal. In Latour, B. and Weibel, P. (eds), *Making things public: Atmospheres of democracy*. MIT Press, Cambridge, MA.

Tarlock, D. and Albrecht, J. (2018). Potential constitutional constraints on the regulation of flood plain development: Three case studies. *Journal of Flood Risk Management*, 11, 48–55.

Thieken, A. H., Petrow, T., Kreibich, H. and Merz, B. (2006). Insurability and mitigation of flood losses in private households in Germany. *Risk Analysis*, 26(2), 383–395.

UVPG (2010). Gesetz über die Umweltverträglichkeitsprüfung. *Federal Law Gazette*, I 94.

VwVfG (2003). Verwaltungsverfahrensgesetz. *Federal Law Gazette*, I 102.

Waldron, J. (2012). *The rule of law and the measure of property*, vol. 63. Cambridge University Press, Cambridge.

WHG (2009). Gesetz zur Ordnung des Wasserhaushalts (Wasserhaushaltsgesetz). *Federal Law Gazette*, I 2585.

Wynne, B. (1996). May the sheep safely graze? A reflexive view of the expert-lay knowledge divide. In Lash, S., Szerszynski, B. and Wynne, B. (eds), *Risk, environment and modernity: Towards a new ecology*. SAGE, London.

3

THE CHANGING NATURE OF FINANCING FLOOD DAMAGES IN CANADA

Heather Bond

Introduction

Flooding is the most frequent and costly natural disaster affecting Canadians today (MMM Group Ltd., 2014). Between 1900 and 1997, 168 significant flood events occurred, predominantly in the southern regions of the country, which are the most populated (Brooks et al., 2001). These events impacted hundreds of thousands of people and cost billions of dollars in damages (MMM Group Ltd., 2014). In addition to the immediate threat, current trends towards climate change and population increases have predicted an increased number of flood events and damage costs in the future (Cherqui et al., 2015). With the likely increasing flood damage and associated expenses, it is necessary to examine critically the current state of Canadian flood management, including the cost recovery mechanisms, to determine the best future options.

Until very recently, the cost recovery scheme following a major flood event in Canada has solely relied on government financial aid to compensate flood victims. Trends towards increasing numbers of flooding events and floods of greater destruction have raised questions as to whether government aid alone is the best option for providing Canadians with financial protection from such disasters. Through examination of the international context, Canada can be shown to be the only G8 country in 2014 not to offer residential flood insurance coverage (Government of Canada, 2014); thus, flood insurance has remained unavailable to Canadian homeowners. However, the landscape for flood management in Canada is shifting, with risk sharing increasingly distributed across homeowners with the recent introduction of private flood insurance in 2015 and 2017. With such a recent policy shift placing greater responsibility on homeowners, the question remains whether this shift will indeed improve managing flood risk in the country.

As Oulahen (2015) points out, there is a gap in our knowledge of the implications of introducing private residential flood insurance in Canada. This chapter aims to investigate these implications from the perspectives of key stakeholders. Options for cost recovery mechanisms are then reviewed, including examples from the international context to highlight challenges and opportunities experienced by other systems.

Context

An important element of flood risk management involves spreading and distributing the financial losses that occur during flooding events. This can be accomplished through private insurance coverage as well as with government aid. As of 2015 in Canada, financial recovery support from flooding events, if covered, was financed almost entirely by provincial and federal governments out of general taxation (Thistlethwaite and Feltmate, 2013). These governments also fund structural defences such as dams and levees (Sandink et al., 2010). The Office of Critical Infrastructure Protection and Emergency Preparedness (OCIPEP) is the overarching federal flood risk management programme that coordinates with the provinces to administer flood relief to the public (EPC, 1999). It is under this programme that the federal Disaster Recovery Financial Assistance Arrangements programme (DFAA) – which covers certain damages to restore public works and repair personal property following severe flood disasters – is administered (EPC, 1999). The DFAA programme is based on cooperation with the provincial governments around a per-capita cost-sharing formula, with the first dollar per person in the province financed by the provincial government and, as damages increase above this level, the proportion of federal assistance increases proportionately (Oulahen, 2015).

However, only severe disasters are deemed necessary for financial aid under DFAA, which classifies a severe flood as one whose damages exceed the CAD$1 per-capita threshold (Sandink et al., 2010). This indicates that major flooding events have been partially covered by government aid and taxpayer funds in Canada, but smaller disasters were self-financed by homeowners. This method of financial flood risk management has worked since its introduction in the 1970s, but increased damages from floods have raised questions from both government and homeowners as to whether it is the best strategy. Chapter 13 demonstrates some of the long-term issues associated with government support of flooded communities and displaced households in Manitoba, Canada.

Floods are becoming a greater problem in Canada due to more extreme weather patterns, increasing population and increasing asset values (IBC, 2015). Adding to the risk is that an estimated 10 per cent of the country's population live in high-risk flood zones, such as floodplains, urban areas with inadequate stormwater drainage, or in low-lying coastal areas (Hodgson, 2018). Furthermore, areas that were previously identified as low risk are now also seen as susceptible due to the increased frequency and strength of precipitation in the form of intense micro-burst rain events (Water Canada, 2017). Climate change is also causing more mid-winter warm spells which can create ice jam hazards such as the flood event in southern

Ontario in February 2018 that originated from the release of an ice jam blocking the Grand River (Semeniuk, 2018).

However, the pivotal event in Canada's recent history was the southern Alberta flood of 2013, referred to as catastrophic and one of the largest natural disasters in Canadian history (Thistlethwaite and Feltmate, 2013), causing damages of up to $5 billion dollars (Lamphier, 2013). To meet the demand for financial assistance following such floods, the annual federal payments covering flood damages have more than quadrupled since 1970 (IBC, 2015). Average annual payments in 2012/13 were $280 million (PSC, 2015), with $3.7 billion paid between 2010 and 2014 following the major events in Alberta and Toronto in 2013 (IBC, 2015). Media coverage of the 2013 summer floods in southern Alberta and the Greater Toronto Area prompted national discussion of the issue (Oulahen, 2015). The experience from the 2013 floods in Canada led the public, insurance companies and the government to examine more closely the policies surrounding financial aid following flood disasters so as to make improvements to residential homeowner satisfaction and overall equity.

With government aid funded through public taxes as the main form of financial relief, the taxpayers of Canada have the burden of reimbursement for flood losses. There are several drawbacks with such flood risk management. These drawbacks include the lack of incentives for the development of individual homeowner mitigation measures and flood hazard maps, as well as the unfairness for taxpayers to finance flooded areas not connected to them (Paudel, 2012). The predictions for increased flooding events and subsequent damage in the future mean greater taxpayer spending on certain parts of the country that are prone to flooding. Additionally, reliance on government aid can be inefficient and can prevent the reduction of development in flood-prone areas (Anderson, 2000).

To address these issues, greater consideration has been given to the Canadian insurance industry financing flood damages. As of 2015, the government funding of financial relief from floods occurred because comprehensive private flood insurance did not exist in Canada. Until then, insurance companies made an important distinction regarding the source of water damage from flood events. It was possible for homeowners to add policies through certain insurance firms that covered water damage caused by sewer backups into properties, but this would not cover 'overland flooding' (Oulahen, 2015). Overland flooding is defined by the Insurance Bureau of Canada (2000) as water that enters a residential home through windows, doors and cracks. This distinction was made to ensure that insurance companies were not responsible for financing flood damages, as providing that coverage was deemed economically unviable for insurers (Thistlethwaite et al., 2017). So, while some home insurance policies covered water damage from sewer backups, they did not cover aboveground water damage.

There are advantages to the insurance industry in expanding their coverage to include overland flood insurance, as it is a missed business opportunity, a major source of claims and a reputational problem (Oulahen, 2015). As of 2015, water damage was in any case a major financial component of insurance companies in

Canada, despite policies which excluded damage caused by overland flooding. In a 2014 report by Aviva, one of the country's largest insurance companies, the principal source of residential property claims was water damage, surpassing both fire- and theft-related claims.

Several possible explanations could account for these high claims from water damage. First, it can be difficult to uncover unambiguously the cause of water damage in homes (Sandink et al., 2010). In addition, the risk of reputational damage has prompted companies to cover claims for flood damage, despite the clear omission of it within insurance policies. The reputational risk stems from a false sense of flood insurance coverage held by the majority of homeowners. Sandink et al. (2010) conducted a country-wide study of 2,100 Canadian homeowners and found that 70 per cent falsely believed that damages caused by overland flooding were covered in their insurance policies. When homeowners discovered they were not covered for flood damages during times of financial crisis following a flood, they were often shocked and angry, leading to general dissatisfaction with insurance companies. This anger and dissatisfaction manifested itself in community groups of flood victims coordinating to 'name and shame' companies that refused to cover damages (Thistlethwaite et al., 2017). In a competitive Canadian insurance industry of over 210 firms, such customer dissatisfaction has a strong impact on individual companies' reputation (Oulahen, 2015; Thistlethwaite et al., 2017). Because of the reputational risk, insurers determined in some instances that it was in the best interest of the company to pay out claims for damage caused by overland flooding irrespective of their true liability (IBC, 2015).

The insurance industry reacted in 2015 to this situation, with the introduction by three major companies of private residential flood insurance coverage. Aviva Canada, The Co-operators and RSA Canada had all introduced some form of overland flood insurance by November 2015 (FloodList, 2015). As of 2017, many more insurers were offering optional add-on overland flood cover (IBC, 2017) and full market adoption is considered to be just 'a matter of time' (Stelmakowich, 2013, 1). The nature of financial flood relief is therefore changing in Canada, and it is important to understand the perspectives of the main stakeholders during this transition.

Attitudes of stakeholders

In examining the attitudes of stakeholders in the Canadian flood-financing environment, it is sensible to understand the objective principles behind the introduction of insurance policies generally. Arnell (2000) describes five criteria that should be met before an insurance company would want to partake in any new form of coverage:

1. It is possible to estimate the likelihood of losses through risk assessments
2. Individual property loss should be independent of other property losses, not all impacted by one event

3. Damaging events occur by chance, randomly and cannot be predicted
4. A sufficient number of people seek insurance coverage
5. The premium should be affordable.

Flood damages inherently do not meet many of these basic conditions. Flooding events often affect large geographical areas and many properties (Arnell, 2000). They are not random and are prone to occur in floodplain areas and high-risk zones, so condition three of unpredictability is not met (IBC, 2015). Finally, flood damages may not meet criteria four or five because of adverse selection, as typically only those homeowners who are most exposed to the hazard of floods will purchase coverage if that purchase is not mandatory.

Building on Arnell's criteria, a study conducted by Thistlethwaite and Feltmate (2013) determined that for economic viability of overland flood insurance by Canadian home insurers, the following conditions must be met:

1. An insurer must be able to accurately price the probability of a flood event occurring, and the losses generated by the event.
2. Premiums can be priced at an affordable level but compensate the insurer for its costs.
3. Premiums are priced at a level that ensures a modest profit to compensate insurers for the additional assumed risk.
4. Premiums should encourage the adoption of risk-mitigation measures by policyholders. Insurers have expressed concern about 'moral hazard' – when policyholders forgo investments in mitigation if they are covered for damages.

These conditions should be considered while reviewing the following sections on the attitudes of government, insurance companies and public towards the introduction of private insurance as the cost recovery scheme following major flood events in Canada.

Government

The government has historically played a key role in the financial relief of flood damages in Canada, through an arrangement of municipal, provincial and federal government provisions.

The federal government attitude towards these provisions was given in the 2014 budget report, which acknowledged the issue of inadequacies in current protection and cost recovery from overland flooding events (Government of Canada, 2014). The report also announced the intention of the federal government to consult with the insurance industry concerning its role in providing private flood insurance to homeowners (Government of Canada, 2014). This attitude has become stronger, with a key objective of the 2015 National Disaster Mitigation Program to work to facilitate private insurance for overland flooding (PSC, 2017).

These announcements indicate that the federal government is dissatisfied with the previous government's aid approach to financing flood damages and supports other options, particularly the involvement of private industry.

Insurers

As the providers of residential overland flood insurance, it is important to understand the perspectives of the insurance industry, taking into consideration the difficulty of comprehending an industry encompassing over 210 insurers. A University of Waterloo study by Thistlethwaite and Feltmate (2013) interviewed senior executives of the largest insurance companies in Canada and found these insurers were divided as to whether the introduction of residential overland flood insurance was viable for their companies. Despite this level of disagreement, almost all insurers did agree that the government emergency relief system for flood damages was reactive and unsustainable.

Although there is no consensus among the insurance industry concerning the viability of flood insurance, there are common concerns shared about its introduction. Most notably, many Canadian insurers are deterred because flood risk maps, initiated by the federal government's Flood Damage Reduction Program in 1975, are outdated and inaccurate (Oulahen, 2015). In fact, 'poor availability of reliable flood plain maps' was the most significant risk identified by these Canadian insurers, making it very difficult to accurately assess risks and price premiums (see Table 3.1; Thistlethwaite and Feltmate 2013, 9). The nature of existing flood maps is as 'hazard' maps which are useful for land use planning by government but do not provide enough information needed by insurers such as on the degree of flood probability (Sanders et al., 2005). Insurers are therefore unable to meet the first conditions identified by both Arnell (2000) and Thistlethwaite and Feltmate (2013).

Further concerns were identified by most insurers, including uncertain attitudes towards flood insurance introduction by the government and public (Thistlethwaite and Feltmate, 2013): clear cooperation with the government is necessary as it will be funding and providing mitigation measures needed to reduce risk to insurers. Concerning public opinion, insurers need to be clear that public demand to purchase the insurance premiums on offer will be widespread and sufficient (Arnell, 2000). Public demand could be identified through willingness-to-pay studies. However, without the cooperation of public and government and the development of reliable flood risk assessments, insurers felt that it would not be reasonable to introduce overland flood insurance. An industry consensus was not considered necessary for insurance introduction (Thistlethwaite and Feltmate, 2013).

This was proven in 2015 because although there was no industry consensus, Aviva Canada, The Co-operators and RSA Canada introduced their overland flood insurance policies (FloodList, 2015). The above criteria had not fully been met, but pressure from public and government stakeholders influenced the decision to introduce this cover (Thistlethwaite et al., 2017). Thistlethwaite and Feltmate

TABLE 3.1 Level of consensus among insurance companies interviewed, concerning key risks and opportunities with introducing overland flood insurance (modified from Thistlethwaite and Feltmate, 2013, 38).

Risks	Level of consensus
Flood mapping data is insufficient	High
Adverse selection impacts on profitability	Low
Moral hazard	Low
Reputational and regulatory risks linked with high costs	Low
Climate change could increase flooding risks	High
Opportunities	Level of consensus
Anticipate customer demand for enhanced coverage	High
Reduce exposure to reputation and regulatory risk	Low
Generate additional source of revenue	Low

(2013) found that the most positive consequence perceived by insurers through the introduction of this overland flood insurance was maintaining a positive brand image through anticipating customer needs (see Table 3.1), important in the highly competitive insurance industry in Canada (Oulahen, 2015). Other benefits identified by insurers were of decreasing reputational and regulatory risks and gaining an alternative source of revenue income (Thistlethwaite and Feltmate, 2013).

While there are therefore key reservations about the adoption of overland flood insurance by Canadian insurance companies, they also note beneficial opportunities and these seem to have outweighed the reservations surrounding not meeting the identified conditions set out in Thistlethwaite and Feltmate's (2013) and Arnell's (2000) criteria. As of 2018, the flood insurance market in Canada is still new, but most major companies now offer some form of overland flood coverage; however, coverage is uneven and properties in high-risk flood zones are still very expensive to insure (Hodgson, 2018).

The public

Another key stakeholder group here are homeowners and the public generally. Homeowners are both the customers of insurers for flood insurance plans and they are the constituency of the federal government concerned with paying taxes and thereby providing financial relief to flooded zones. International studies have found that a number of factors are significantly correlated with public demand and their purchase of flood insurance, including a homeowner's perceived risk, previous experience with flooding, the price of insurance premiums and income level (Kousky, 2011; Lamond et al., 2009). As Oulahen (2015) found in a survey conducted on Vancouver residents, demand for flood insurance is entwined with a complex set of determinants of vulnerability.

Elaborating on the findings of Oulahen (2015) following the introduction of overland flood insurance coverage by the three companies in Canada, a public opinion survey was developed in 2016 by researchers at the University of Waterloo (Henstra and Thistlethwaite, 2017). This survey gathered data from 2,300 Canadians in all ten provinces, to provide a better understanding of the current public perception towards flood insurance and damage relief in Canada, targeted specifically at citizens living in high flood risk areas (Henstra and Thistlethwaite, 2017). These survey participants were asked 57 questions regarding their perceived flood risk and actions to protect against potential damage from flood events.

The results of this survey revealed several key findings on public opinion regarding flood risk and protection in Canada. First, there is a clear misalignment between real and perceived flood risk among the public. Only 6 per cent of the respondents knew that they were living in high-risk areas, with 50 per cent expressing no concern at all about floods (Thistlethwaite et al., 2017). Furthermore, 79 per cent of respondents did not believe flood risk was likely to increase in the next 25 years. This demonstrates the worryingly low level of flood risk awareness by Canadians living in such high-risk flood zones.

Concerning responsibility for flood risk management and financing, the survey showed that Canadians are open to sharing responsibility, and consider that insurance companies should finance flood damages. More than 80 per cent of respondents felt they should share some of the responsibility to protect their property (Thistlethwaite et al., 2017). Furthermore, the survey results suggest government should continue to be actively involved in flood risk management.

Responses from this survey also demonstrate that options for financing flood damages in Canada remain unclear to many residents. Most of the 2,300 respondents either believed or were uncertain as to whether their home insurance policy covered damages from overland flooding (Henstra and Thistlethwaite, 2017). This is similar to the findings from the national survey by Sandink et al. in 2010, in which many participants falsely believed their insurance policies covered all flood damages. This indicates that there may be little change in public awareness since 2010, despite major flooding events and the introduction of private flood insurance coverage. As such, this level of public awareness needs purposeful raising in future flood risk management strategies.

These results underline a lack of motivation for further seeking private flood insurance. Indeed, only 23 per cent of respondents expressed an interest in purchasing overland flood cover, and most were not willing to pay more than CAD $100 annually, a premium that is not economically viable for private insurance companies to offer homeowners in high-risk areas (Henstra and Thistlethwaite, 2017). This is of great concern for two reasons. The slack demand makes it difficult for home insurance companies to provide premiums for overland flooding at affordable prices if cover remains voluntary. Additionally, as the federal and provincial governments in Canada begin to distance themselves from financial aid relief, the burden falls on individual homeowners to pay themselves for the damages incurred when disasters strike.

A review of insurance options

A review of the international context shows that other G8 countries have also faced struggles with national flood risk management. While no one country's solution offers a template for Canada's financial flood relief strategies moving forward, they provide a range of options (Table 3.2). Each option has its benefits and drawbacks and reviewing these suggests which directions would relate best to the Canadian system.

Public flood insurance models are used by the United States and France, with taxpayers financing damages. The National Flood Insurance Program (NFIP) conducted through federal, state and local governments in the United States is an optional insurance coverage that has led to adverse selection as well as major government subsidy of many high-risk areas (IBC, 2015). Thus, NFIP has been deemed financially unsustainable (Sandink et al., 2016). Chapter 15 in this book explores some of the issues with NFIP more closely, including its general insolvency. In France, flood insurance is included in standard mandatory home insurance policies, making penetration nearly universal (IBC, 2015), with all policyholders charged a single rate, regardless of their level of risk (IBC, 2015). The problem of adverse selection is removed through the mandatory nature, but Lamond and Penning-Rowsell (2014) describe how providing a set rate removes homeowner incentives to mitigate risk.

TABLE 3.2 Summarised comparison of international flood insurance programmes of G8 countries, excluding Canada (modified from IBC, 2015, 13)

	Model	*Purchase*	*Packaging*	*Take-up*	*Pricing*	*Financial impact mainly borne by*
France	Public	Mandatory	Bundled (with other catastrophes)	100%	Government-set	Taxpayers
USA	Public	Mandatory	Optional (add-on)	20–30%	Government-set	Taxpayers
Germany	Private	Voluntary	Optional (add-on)	25–30%	Risk-based	Policyholders
Italy	Private	Voluntary	Optional (add-on)	<10%	Risk-based	Taxpayers
Russia	Private	Voluntary	Optional (add-on)	<5%	Risk-based	–
Japan	Private	Voluntary	Bundled (with comprehensive home-owners policy)	40%	Risk-based	Policyholders
UK	Private	Voluntary	Bundled (with homeowners policy)	95%	Risk-based	Policyholders

As Table 3.2 outlines, many G8 countries have a voluntary private flood insurance model that is risk-based. Japan and the United Kingdom offer these models with bundled insurance packages bought voluntarily (IBC, 2015). This has worked for the UK under an informal 'Gentleman's agreement' made in 1961 between insurers and the government, making insurance nearly universally available to all but the highest-risk properties, provided the government funds sufficient flood defences and hazard maps (Sandink et al., 2016). To cover high-risk properties with otherwise high premiums, a risk-sharing pool called Flood Re is being developed by insurers and government (Defra, 2014; Penning-Rowsell and Priest, 2015). Germany's private model is similar, with the key distinction that flood insurance is an optional add-on for homeowners (IBC, 2015). The effect of this may be seen in the difference in insurance penetration by residents in the UK and Germany, as German flood insurance has failed to cover most residents. Optional flood insurance in Canada could experience strong adverse selection, and very low up-take as in Germany, Italy and Russia, if not bundled into other forms of property cover. Chapter 2 explains some of the reasons for poor insurance uptake in Germany.

These international examples shed light on other methods of national flood insurance programs that could be applied in Canada. Sandink et al. (2010) asserts that the UK model offers the best framework for Canada, due to its high penetration rates and recent attempts to cover high-risk areas with Flood Re. There are still challenges with this model, however, including the possibility of a lack of government investment, a reduced affordability of home insurance through the bundling of policies leading to more cover than the public will pay for, and reduced communication of flood risk to homeowners through price signalling. Thistlethwaite and Feltmate (2013) suggest a strategy of optional flood insurance to address this issue, although this revisits the problem of adverse selection. Despite this, we favour optional private insurance in Canada as being easiest to implement in the current context and alignment with the attitudes of the government, insurers and the public. A long-term option to later include bundled flood insurance in homeowner packages would aid the rate of uptake and reduce premiums.

Conclusions

Canada is facing a changing landscape of flood risk and financial recovery, and it is difficult to identify the best strategy to address this.

Historically in Canada, property owners have largely recovered flood damages from the government, with the cost being spread over all taxpayers in the country through the DFAA programme. The federal government and insurance industry generally agree that this reliance on government aid to recover flood damages is unsatisfactory and unsustainable. Canadians think the private insurance industry should bear the responsibility for financial recovery but show little interest and willingness to pay for comprehensive flood insurance themselves, possibly due to past reliance on government aid and a lack of knowledge of the subject.

While home insurance companies begin to offer overland flood protection, it will take several years before affordable premiums are available to most Canadians (Water Canada, 2017). A global review demonstrates that there are several options to finance flood damages through insurance, with optional private insurance offering an attractive and feasible choice in Canada.

However, there are a number of general conditions identified by the IBC (2015), which need to be in place to provide the strongest flood risk management scheme. It is most important to have updated flood risk maps and limited reliance on government financial aid to incentivise homeowners to take mitigation measures and prevent taxpayers from large burdensome expenses. Furthermore, adequate funding of structural flood defences is also necessary to protect homeowners, as well as engaging and informing the public on their flood risk and available insurance options.

References

Anderson, D. R. (2000). Catastrophe insurance and compensation: Remembering basic principles. Society of Chartered Property and Casualty Underwriters. *CPCU Journal*, 53 (2), 76–89.

Arnell, N. (2000). Flood insurance. In D. J. Parker (ed.), *Floods*. London: Routledge.

Aviva (2014). Over 50% of all home insurance claims caused by water damage in 2013, Aviva Canada data shows. [Online] Available at www.newswire.ca/news-releases/over-50-of-all-home-insurance-claims-caused-by-water-damage-in-2013-aviva-canada-data-shows-513988141.html. [Accessed 28 Jul 2018]

Brooks, G. R., Evans, S. G. and Clague, J. J. (2001). Flooding: A synthesis of natural geological hazards in Canada. *Geological Survey of Canada Bulletin*, 548, 101–143.

Cherqui, F., Belmeziti, A., Granger, D., Sourdril, A. and Le Gauffre, P. (2015). Assessing urban potential flooding risk and identifying effective risk-reduction measures. *Science of the Total Environment*, 514, 418–425.

Defra (Department for Environment, Food and Rural Affairs) (2014). A short guide to Flood Re. Defra, London.

EPC (Emergency Preparedness Canada). (1999). *Federal Disaster Assistance Arrangements (DFAA)*. Emergency Preparedness Canada, Ottawa.

FloodList (2015). Canadian flood insurance now available from RSA, Aviva and The Co-operators. [Online] Available at: http://floodlist.com/america/canada-flood-insurance-rsa-aviva-cooperators. [Accessed 28 Jul 2018]

Government of Canada. (2014). The road to balance: Creating jobs and opportunities. 2014 Federal Budget. Government of Canada, Ottawa, 427.

Henstra, D. and Thistlethwaite, J. (2017). Flood risk and shared responsibility in Canada: Operating on flawed assumptions? Policy Brief No. 116. Centre for International Governance Innovation, Waterloo.

Hodgson, G. (2018). How can we better mitigate flood risk in Canada? [Online] Toronto, The Globe and Mail Inc. Available at: www.theglobeandmail.com/report-on-business/rob-commentary/how-we-can-better-mitigate-flood-risk-in-canada/article37609942/. [Accessed 20 Apr 2018]

IBC (Insurance Bureau of Canada) (2015). The financial management of flood risk. An international review: Lessons learned from flood management programs in G8 countries. IBC, Toronto.

IBC (Insurance Bureau of Canada) (2017). Water. [Online] Available at: www.ibc.ca/qc/disaster/water. [Accessed 30 Apr 2018].

Kousky, C. (2011). Understanding the demand for flood insurance. *Natural Hazards Review*, 12, 96–110.

Lamond, J., Proverbs, D. and Hammond, F. (2009). Accessibility of flood risk insurance in the UK: confusion competition and complacency. *Journal of Risk Research*, 12(5), 825–840.

Lamond, J. and Penning-Rowsell, E. (2014). The robustness of flood insurance regimes given changing risk resulting from climate change. *Climate Risk Management*, 2, 1–10.

Lamphier, G. (2013). Alberta flood tab could run as high as $5B: Too early for insurer of 278 municipalities to estimate costs. [Online] *Edmonton Journal*. Available at: www2.canada.com/edmontonjournal/news/business/story.html?id=b089dd83-4d5a-401e-bbb9-2a10c51593e2. [Accessed 10 Jan 2016]

MMM Group Ltd. (2014). National floodplain mapping assessment: Final report. Public Safety Canada, Ottawa, 68.

Oulahen, G. (2015). Flood insurance in Canada: Implications for flood management and residential vulnerability to flood hazards. *Environmental Management*, 55(3), 603–615.

Paudel, Y. (2012). A comparative study of public–private catastrophe insurance systems: Lessons from current practices. *The Geneva Papers on Risk and Insurance: Issues and Practice*, 37(2), 257–285.

Penning-Rowsell, E. C. and Priest, S. (2015). Sharing the burden of increasing flood risk: Who pays for flood insurance and flood risk management in the United Kingdom. *Mitigation and Adaptation Strategies for Global Change*, 20(6), 991–1009.

PSC (Public Safety Canada) (2015). The government of Canada announces disaster mitigation investments and modernizes disaster financial assistance arrangements. [Online] Government of Canada, Ottawa. Available at: http://news.gc.ca/web/article-en.do?nid=922139. [Accessed 14 Jan 2016]

PSC (Public Safety Canada) (2017). National Disaster Mitigation Program (NDMP). [Online] Available at: www.publicsafety.gc.ca/cnt/mrgnc-mngmnt/dsstr-prvntnmtgtn/ndmp/index-en.aspx. [Accessed 28 Apr 2018]

Sanders, S., Shaw, F., MacKay, H., Galy, H. and Foote, M. (2005). National flood modelling for insurance purposes: Using IFSAR for flood risk estimation in Europe. *Hydrology and Earth System Sciences Discussions*, 9(4), 449–456.

Sandink, D., Kovacs, P., Oulehan, G. and McGillivray, G. (2010). Making flood insurable for Canadian homeowners: A discussion paper. Institute for Catastrophic Loss Reduction/Swiss Reinsurance Company Ltd., Toronto.

Sandink, D., Kovacs, P., Oulahen, G. and Shrubsole, D. (2016). Public relief and insurance for residential flood losses in Canada: Current status and commentary. *Canadian Water Resources Journal*, 41(1–2), 220–237.

Semeniuk, I. (2018). Winter flooding risk in Canada expected to increase as climate warms. [Online] Toronto, The Globe and Mail Inc. Available at: www.theglobeandmail.com/technology/science/winter-flooding-risk-in-canada-expected-to-increase-as-climate-warms/article38078061/ [Accessed 20 Apr 2018].

Stelmakowich, A. (2013). Overland overhaul. *Canadian Underwriter*. [Online] Available at: www.canadianunderwriter.ca/news/overland-overhaul/1002551174/. [Accessed 15 Jan 2016]

Thistlethwaite, J. and Feltmate, B. (2013). *Assessing the viability of overland flood insurance: The Canadian residential property market*. The Co-operators Group Ltd/University of Waterloo, Waterloo.

Thistlethwaite, J., Henstra, D., Peddle, S. and Scott, D. (2017). Canadian voices on changing flood risk: Findings from a national survey. [Online] Interdisciplinary Centre on Climate Change and Partners for Action (P4A), Waterloo. Available at: https://uwa terloo.ca/climate-centre/news/canadian-voiceschanging-flood-risk-findings-nationa l-survey. [Accessed 28 Apr 2018]

Water Canada (2017). Canada needs a national strategy to address flood risk. [Online] Available at: www.watercanada.net/feature/canada-needs-a-national-strategy-to-address-flood-risk/. [Accessed 30 Apr 2018]

4

POWER FOR CHANGE IN ADAPTING TO COASTAL FLOOD RISK ON CURACAO IN THE CARIBBEAN

Lena Fuldauer

Introduction

The Caribbean Small Island Developing States (CSIDS) are widely acknowledged as being particularly affected by increasing climate impacts (Nurse et al., 2001; Tompkins et al., 2005). Given that 70 per cent of all economic activity takes place within approximately three kilometres of the coastline (Duval, 2004), these islands are especially vulnerable to coastal sea-surges from hurricanes and floods. However, despite their importance, most Caribbean islands remain in the early stages of implementing flood risk management (FRM) initiatives for coastal climate adaptation.

Which flood risk measures are included or excluded is contested and involves trade-offs between risk reduction and economic opportunities, most notably tourism developments (Tunstall et al., 2009). Power relations between stakeholders play a large role in such decision-making, including who or what is considered at risk or which measures for FRM are ultimately prioritised. To date there are few tools available that enable policy makers or researchers to explicitly identify power dynamics related to climate adaptation and reduce the potential harmful effects of such power orientations.

The aim of this chapter is threefold: first, to identify influential actors in coastal FRM initiatives; second, to understand power relations between the most influential actors; and third, to propose strategies to manage these power relations for more effective FRM. In addressing these aims, a novel actor-mapping methodology has been developed and applied to coastal FRM, focusing on the island of Curacao.

Power theories and actor mapping

Over the past decades, many authors have defined power, using different conceptualisations (Weber, 1947; Ailon, 2006; Allen, 2009). According to these ideas, actors can exercise power – synonymous with influence here – by dominating other actors directly or indirectly or through resources other actors value. Following the theoretical underpinning of actor-network theory (Callon, 1986), actors can be considered both human or non-human, with each actor being made up of a network. These second-level actor networks are not assessed here. Non-human actors are defined as actors that influence decisions where no responsible individual can be credited with the decision. This includes broader systems, costs or specific policies that are considered influential in FRM decision-making. In understanding why and how certain actors influence others, an analysis of power theories is warranted. This chapter focuses on four main theories of power (Table 4.1), which have been chosen for their ability to be applied to stakeholder actor-mapping. The first three theories build upon each other and view one actor as dominating, while the last one considers both actors as having power simultaneously.

Power dynamics are commonly visualised through social network analysis (Knoke and Yang, 2008), with stakeholder influence diagrams using 'power versus interest grids' (Bryson, 2004) or star diagrams (Eden and Ackermann, 2013). However, these approaches share a number of limitations. First, they are prone to what Matthews (2008) calls the 'ontological and teleological dilemma'. While the former suggests that systems mapping is limited by narrowly defined boundaries that ignore unintended consequences, the latter implies that even when adopting a holistic systems view, improvements from all agents cannot be guaranteed due to fundamental value differences. Second, they focus on mapping human actors only, leaving out non-human actors, which can significantly influence decisions. Third, these methodologies define power as centrality, the quantity of network links associated with a given actor (Knoke and Yang, 2008). This inherently places a

TABLE 4.1 Overview of selected power theories used in this chapter

Author	Theory	Explanation
Dahl (1957)	Overt dimension: power as decision-making	Actor 'A has power over B to the extent that he can get B to do something that B would not otherwise do'
Bachrach and Baratz (1962)	Covert dimension: power as non-decision-making	Non-decision-making includes 'A not listening to B's demands or B not raising a problem with A in the belief that A may disagree, hence impeding decision-making on issues important to B'.
Lukes (2004)	Latent dimension: real interests and subconscious decision-making	'A may exercise power over B by getting him to do what he does not want to do, but he also exercises power over him by influencing, shaping or determining his very wants'.
Emerson (1962)	Power as resource dependency	Power is determined by an actor's ability to control resources or needs the other actor values: 'A depends upon B if he aspires to goals or gratifications whose achievement is facilitated by appropriate actions on B's part'.

focus on visible relationships (Dahl's overt dimension), thereby ignoring other power dimensions (Table 4.1).

Notably, multi-level stakeholder mapping (MLSM) is a power-elucidating technique that aims to address the first limitation (Mayers and Vermeulen, 2005). By inviting participants from multiple levels, MLSM is considered more effective in assessing the relative influence of actors when compared to central-level mapping (Sova et al., 2015). However, to date, no methodology exists which uses power dimensions to select actors and analyse actor dynamics or which integrates non-human actors in such multi-level stakeholder mapping. The research reported here builds upon MLSM by analysing dynamics between actors and grounding such an analysis in power theories. A three-step methodology for a hybrid mapping technique (see Figure 4.1) is developed, termed multi-level power dynamics mapping (MLPDM). This gives a snapshot of currently influential human and non-human actors and their power dynamics. Its application can pave the way for more pluralistic analyses of the sources of power and identify effective system changes.

Methodology and methods

Case study selection and data collection

Curacao was chosen as a CSIDS case study for a number of reasons. First, contacts to decision-makers were already established in previous research, which facilitated data collection. Second, Curacao has a relatively high coastal population and tourist density, with 150,000 residents and 700,000 yearly tourists (The Ministry of Health Environment and Nature, 2014). Lastly, Curacao has been considered a Caribbean island at risk of coastal climate-induced flooding (Simpson et al., 2010).

Mixed methods included a quantitative risk assessment, a qualitative content analysis, and interviews in November and December 2017. An initial assessment of coastal flood risk was performed using elevation data from NASA (2017) and available sea-surge projections (Muis et al., 2016). Tourism properties at highest risk of sea-surges were contacted as potential interviewees. Furthermore, a content analysis of primary and secondary sources was performed, including government press releases, reports and organisational websites.

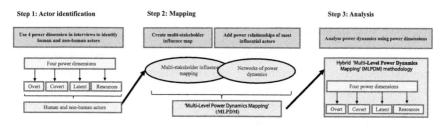

FIGURE 4.1 Steps in the hybrid methodology developed in this research: multi-level power dynamics mapping (MLPDM)

TABLE 4.2 Interviewees (conducted in December 2017)

Organisation type
Government fire department
Government meteorological department
Government urban planning department
Academia
Non-Governmental Organisation (NGO)
Tourism sector – international hotel
Tourism sector – local, small hotel
Tourism sector – airport
Local NGO
Civil society

From this analysis, possible governmental and local participants were identified, who were asked to suggest further stakeholders concerned or affected by coastal flood risk, using non-probability 'snowballing' methods to determine informal networks not discernible from published material (Tansey, 2007). From 21 possible participants, ten were chosen for interviews based on the diversity of level they represented (local, national, international) (Table 4.2). This helped to combat critiques of power mapping whereby focus group composition consist of central-level interviewees only. Nevertheless, it should be noted that results from this research are based on information from stakeholders available at the time of the research period. Notable is the absence of a response from the department of risk reduction.

Semi-structured interviews were conducted via Skype with participants' consent and cooperation in line with Oxford Ethics Committee guidelines. Notes were taken and analysed, and participants were later anonymised. Following the approach of Sova et al. (2015), care was taken to create an environment where respondents felt comfortable in sharing their perspectives. Following step 1 of our methodology (Figure 4.1), interviewees were asked to identify actors in the light of different power dimensions, describe their relative influence and explain relationships between actors, excluding themselves.

Mapping and data analysis

In line with steps 2 and 3 of our methodology (Figure 4.1), a multi-level stakeholder map was first created from interviewee responses indicating the perceived influence of each actor group. Actors were only included in the map if considered influential by at least three interviewees. Second, power dynamics between the most influential proponents and opponents of coastal FRM were analysed to create the multi-level power dynamics map. Perception-based data was used for the power analysis of this chapter, which is justified as it is also at the heart of bias formation in FRM.

Results

Influential actors in coastal FRM on Curacao

With regard to coastal FRM proponents, the most influential actors include the Ministry of Traffic, Transport and Urban Planning (MTTUP)'s fire and meteorological department. Respondents cited the departments' respective mandates for 'ensuring safety' and 'watching the weather to protect life and property' (Meteorological Department Curacao, 2017), its expertise in the area, its extensive network and motivated employees as key factors of influence.

The tourism sector, particularly coastal high-end resort hotels and the airport, is considered nearly as influential due to its financial resources and its incentive to continue attracting tourists. Local NGOs are described as implementers without which there would be less action and considered influential due to their local knowledge, large networks and lack of political or economic motivations. These can be considered bridging agents as they are identified as influential by interviewees operating at different actor levels within Curacao.

Still influential, but to a lesser degree than the above-mentioned actors, are regional and international organisations. The World Bank, Caribbean Climate Centre and UN agencies are powerful actors (see Figure 4.2) because of their significant financial resources, expertise in the field and relative autonomy from central government. However, these groups are not considered influential by local stakeholders given that, historically, few of their proposed FRM recommendations have been taken up.

With regard to coastal FRM opposition, the most influential non-human actors include Curacao's political system and its corruption, which deters change from happening. While politicians should arguably respect democratic accountability by adopting policy in line with public values, our interview participants quoted a number of false promises made by Curacao politicians to their voters. This behaviour is incentivised by a system in which politicians, once elected, are secured lifelong benefits from taxpayers' money. As such, the corrupt system inhibits voters to effectively use their decision-making power in politics.

The MTTUP's urban planning department is considered an influential human opponent of FRM due to its mandate on urban planning. However, this mandate has also been used to support FRM initiatives. This suggests that actors have ambivalent roles with regard to their influence in implementing FRM initiatives. Similarly, different entities within the tourism sector are also either proponents, opponents or both. While the tourism sector can support FRM initiatives, it has also been identified as an influential opponent through the destruction of flood-protective mangrove sites for tourism developments. As such, actor groups are not necessarily homogeneous entities but instead are made up of heterogeneous elements. This finding aligns with the framework of actor-network theory, implying that each actor is made up of networks that constantly change and interact (Callon, 1986)

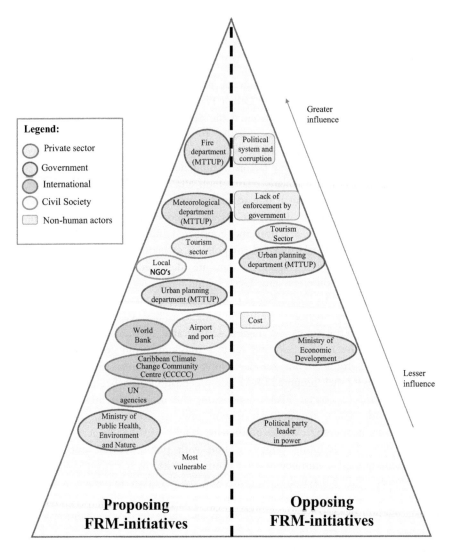

FIGURE 4.2 Multi-stakeholder influence map of Curacao's FRM regime in 2017, following the method of Sova et al. (2015)

Notes: The figure further includes human and non-human actors and differentiates between FRM proponents (left) and opponents (right). The relative size of the circle to the policy apex represents the perceived influence of each actor group.

Equally informative are those actors that have not been identified as influential. This group includes vulnerable communities living near the Curacao oil refinery where floods have previously inflicted severe damage. It suggests Curacao's slow-moving advances in linking environmental and health issues from possible floods with planning and policy-making for effective coastal zone management. The lack of interest from Curacao figures of authority in sustainability measures is documented elsewhere (Dinica, 2006)

Power relationships in coastal FRM in Curacao

An emphasis is placed here on relations with the Curacao tourism sector only, a highly influential – yet often overlooked – group considered at risk of coastal climate impacts (Simpson et al., 2010; Scott et al., 2012).

The relationship between the MTTUP's fire and meteorological departments and the tourism sector is an example of Dahl's view of power as decision-making (see Figure 4.3), as these departments have interacted with hotels, airports and ports in implementing tailored flood warning systems (Meteorological Department Curacao, 2015). Dahl's overt power dimension is also evidenced by the urban planning department's decision to relocate a desalination plant away from its coastal flood-risk location so as to allow development of a tourism boulevard, which is projected to attract tourists and contribute to economic growth. The urban planning department's influence over this decision entailed the unintended consequence of significantly reducing the risk of critical infrastructure failure through a retreat adaptation option that located the facility away from the flood risk.

Bachrach and Baratz's covert dimension of power is evidenced in the relationship between the tourism sector, the urban planning department and local NGOs. The urban planning department ignored FRM recommendations from local NGOs by using their power of non-acting on proposals. Moreover, it refrained from using its power to restrict coastal tourism developments either directly

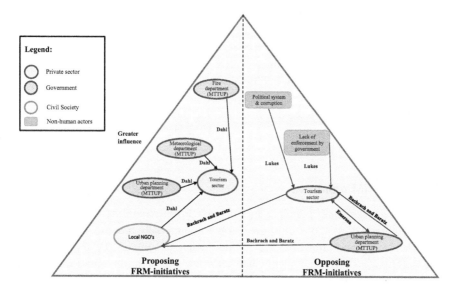

FIGURE 4.3 Multi-level power dynamics map showing power relationships between most influential actors in the FRM regime in Curacao, 2017

Notes: All the arrows ascribed to Dahl denote relations resulting in increased FRM initiatives; all other arrows illustrate power relations which oppose FRM initiatives. Arrows were drawn by applying selected power theories from Table 4.1 to participants' suggestions, published material evidence and recent decisions.

through land use granting permissions or indirectly through not enforcing fines against illegal developments in flood risk areas. These cases are examples of 'B failing to raise an issue with A because B believes A would oppose the idea' (Clegg, 1989, 77).

These decisions can be attributed to contemporary drivers for tourism growth (see Honey and Krantz (2007)) and related resource dependencies. When tourism development applications are denied by the department, the resources associated with such development are not realised. However, these are an indispensable source of government revenue and also create jobs. This example is hence also reflective of Emerson's view of power, which states that both actors can hold power simultaneously.

Lukes' latent power dimension is likely to be present here through institutional nudges and personal influences, such as when politicians favour certain tourism investments over others or elect chosen ministers only. Such institutional nudges can slow the advancement of more stringent FRM initiatives proposed by local NGOs or international agencies, and can be attributed to Curacao's political uncertainty and corruption.

Discussion

The application of the hybrid stakeholder mapping technique in Curacao illustrates how a number of power theories may be at play simultaneously. None fully explains power dynamics, a finding that aligns with previous research (Self and Penning-Rowsell, 2017). Instead, theories are ways of conceptualizing dynamics and in reality many relationships overlap. They contain personalities, may be applicable to different extents at different times, and are often a function of the problem at hand, the policy and the politics at that time (see Kingdon (1984)). Nevertheless, our results provide insights into both the political and complex nature of the FRM regime in Curacao and the practicality of the experimental MLPDM methodology.

Curacao's FRM regime

In analysing stakeholder influence in Curacao's FRM regime, results do not align with traditional power sources in the Caribbean, especially those that are explicitly represented as being influential in published policy documents (Pelling and Uitto, 2001). These include agents with high expertise and large networks such as the central government, UN agencies, the World Bank and the Caribbean Community Climate Change Centre (the CCCCC). While these actors have some influence in Curacao's FRM regime, they are only deemed influential by our respondents within their level of operation, a phenomenon termed 'proximal level bias' (Sova et al., 2017).

Our local-level respondents highlighted that the large networks of central government are often abused, a typical example of the latent power dimension, and not utilised to spark land use planning guidelines. This aligns with the finding by Mycoo and Donovan (2017, 75) that 'the ministries or agencies responsible for these environmental policies often have limited involvement in urban planning and city governance'. In fact, weak government systems and lack of policy enforcement resulting in unauthorised constructions are typical in Caribbean SIDS (Wilkinson, 2004). Our study's power analysis adds to the literature by suggesting that the root cause of such inappropriate legislation and enforcement is related to the latent dimension of power dynamics: the political agenda and true interests of stake-holders, namely short-term economic self-interests.

This is particularly evident in political party leaders, which is also a reason for the relatively little power attributed to central government over FRM here. The complex political agenda of Curacao is shaped by its history. Since the autonomy from the Dutch, the Curacao government has struggled to use its traditional power to achieve public objectives, especially those not directly contributing to economic growth. This is reflective of a popular debate in developing countries over the legitimacy of traditional ruling structures and the reluctance to use traditional power in attaining environmentally related public objectives. For example, Cameron (2011) suggests that politicians are unwilling to own the response agenda to climate change with little domestic account-ability pressure for achievement.

Despite the relatively limited influence of central government, power is attrib-uted to specific, more technical departments of the MTTUP. The fire department's influential role in FRM illustrates Curacao's reactive stance to climate change. However, the increasingly greater influence of the meteorological department over time pinpoints Curacao's vision towards a more proactive stance on climate change, documented elsewhere (The Ministry of Health Environment and Nature, 2014). These technical government departments form the dominant group of elite power holders within the FRM regime. In fact, the creation of such formal departments in the Caribbean has been considered evidence of adaptation (Tomp-kins, 2005).

The private sector and local NGOs are also notably present in this study's power rankings. The high FRM-proponent influence ascribed to the private sector, specifically the tourism sector, is inconsistent with the majority of the wider adaptation discussions to date (Agrawala et al., 2011; Biagini and Miller, 2013; Pauw and Pegels, 2013). In the Caribbean, specifically, the tourism sector has typically been considered an opponent of coastal adaptation (Simpson et al., 2010). Nevertheless, it is well documented that tourism planning involves huge complexities and feedbacks (Dredge and Jenkins, 2007). The high influence of the tourism sector in encouraging FRM initiatives is hence likely reflective of such complexities, including economic interests, and could reduce the latent power of political parties. The finding that local NGOs can act as bridging agents between the private and public sector is consistent with literature on the role of informal networks in driving climate adaptation in Caribbean SIDS (Robinson, 2017).

Managing power relations for more sustainable FRM

That a large, diverse group of actors opposes FRM-initiatives – either overtly, covertly or latently – calls for a system-wide approach to strengthen the power of FRM proponents in favour of risk reduction. This can spur positive feedbacks for climate adaptation and avoid enforcing complex political powers.

Recognising that managing power relations for FRM is context-specific and could involve various stages, strategies and tools, we propose a number of generic steps. First, agents of change should be identified that are considered influential across levels. Second, these agents should ally with FRM proponents to identify initial win-win situations and low-regret options and aim to reformulate arguments from influential FRM opponents. Lastly, appropriate methods, including both material and social elements necessary for long-term adaptation (Buzinde et al., 2010), should be used to scale-up initiatives. In doing so, a framework should be devised to assess the sustainability of different FRM initiatives.

In Curacao, the influential meteorological department could incentivise FRM initiatives by utilizing its technical knowledge, access to international resources and motivated employees, three characteristics of importance for environmental policy implementation in SIDS (Dinica, 2006). Using future windows of opportunity (see Johnson et al. (2005)), the meteorological department could spur change across other government departments. Using the cost argument, which is typically utilised by governments opposing FRM initiatives, alternative economic motivations to that of coastal development can be offered: the saved costs of early adaptation. Securing such early benefits is an effective option discussed in the adaptation literature (Wilby and Dessai, 2010) and could be applied and scaled up in Curacao.

Given the little power attributed to political parties and the absence of enforced government policy in Curacao, private sector initiatives can thrive. The influential economic argument used to build coastal tourism developments could be re-formulated to support hotel or landowner-sponsored development of coastal tourist recreation areas. This may include permeable tourist promenades or seawall facilities acting as tourist sightseeing points. Such examples are win win adaptation options, benefitting both the government through the private sector provision of FRM measures and the tourism industry by increasing the number of tourist facilities. The latter is especially important in light of the trend in the Caribbean towards more cruise ship passengers and shared apartment tourists, who demand recreation areas without increasing the demand for hotel developments. Such strategies also avoid the 'flood defence trap', the notion that landowners are reluctant to implement flood defences because they remind them of flood risks (Harris, 2010).

However, the ability of the tourism sector to implement coastal FRM initiatives depends on their stage in the economic cycle, with resorts in the rejuvenation stage having limited options for coastal defence strategies in light of their size, financial resources and existing built assets (Jennings, 2004). Hence smaller, family-run businesses may be unable to respond to flood risk unless supported. More effective local and regional coordination with appropriate sustainability and feasibility frameworks could help smaller tourist facilities. Consequently – and unlike authors suggesting the need for policy intervention in achieving sustainable tourism (McDavid, 2002; Karagiannis, 2002) – the results of the research reported here imply that a larger influence could be achieved by strengthening tourism and locally led FRM initiatives while reducing input from political parties that have seldom spurred change in the Caribbean.

Local NGOs can champion under-represented voices from the population, including voters, communities vulnerable to flooding impacts and small hotel businesses at risk of losing their livelihoods. They can also advocate for the continued development of mangroves as natural flood defences, which are at risk of destruction from impacts of unplanned coastal tourism (Hall, 2001; Jennings, 2004). Such allocation of discretionary power to local actors is a topic commonly cited in the climate adaptation literature (Dodman and Mitlin, 2013; Granderson, 2014) and could be applied on Curacao. Additionally, NGOs can play a role in raising awareness of the 'self-imposed limitations to growth' (Priestley and Mundet, 1998). In light of Curacao's relatively low hotel occupancy rates, they can advocate to reduce future tourist development while promoting the trend towards shared properties, which is considered a population-inclusive strategy (Clayton and Karagiannis, 2008). Most importantly, local NGOs could render deeply engrained latent dimensions of power evident in order to address the root cause of many social vulnerabilities related to flood risk.

This discussion highlights how an awareness of power dynamics in a specific climate adaptation regime can help reduce the harmful effects of such dynamics and render more sustainable outcomes. We suggest that understanding these power dynamics lies at the heart of achieving broader adaptation goals, such as integrated coastal tourism development. By identifying influential actors that meet the interests of multiple stakeholders and address root problems, coastal climate adaptation is likely to be more effective.

Methodology

The above analysis illustrates how power dynamics evidenced through publications or MSIM alone miss out on the role of non-human actors and latent power dimensions. This includes the political system and underlying priorities and patronages as key barriers to effective FRM.

As such, the application of the hybrid actor mapping technique in the context of coastal FRM reveals that not one of the methods (influence mapping versus power dynamics) merged in this chapter is servant to the other. Instead, they fulfil different purposes: theories of power dynamics helped identify both human and non-human actors and particularly covert and latent power dynamics, while the MSIM provides the analyst with a relative influence of local and central agents – a unique vantage point for the identification of influential bridging agents, under-represented actors and critical disconnects.

Grounding multi-level stakeholder maps in power conceptualisations allowed understanding the ambivalent roles of actors and underlying reasons for such disconnects, particularly those covert or latent ones not otherwise visible. This includes resource relationships, the large influence of the wider political system in opposing FRM initiatives, and the power to ignore recommendations. Such considerations are particularly important in FRM because its effectiveness 'cannot be judged just by what happens in the areas affected: it is what has not happened that counts' (Pottier et al. 2005, 20). This understanding can then be applied to better target recommendations for effective FRM.

Conclusions

By elucidating the underlying structures of decision-making and revealing influential, non-traditional human and non-human actors, power dynamics mapping can provide a basis for allowing more pluralistic and effective flood risk management.

In Curacao, potential bridging actors with high local-level influence such as environmental NGOs could be better engaged at high levels of operation, both in government and the tourism sector, to improve understanding of and response to coastal flood risk. Local NGOs and large hotels themselves – together with environmentally motivated employees in government ministries – can have an important role in encouraging FRM initiatives without the need for input from political parties. This requires increased coordination between formal and informal institutions and between the public and private sector for a common vision on sustainable FRM. Achieving increased pluralism more broadly in the climate change adaptation regime could offer increased responsiveness, legitimacy and transparency in CSIDS, thereby diversifying potential adaptation responses.

The results of this research only provide a snapshot of the power relations at a specific time by selected stakeholders and are prone to common methodological limitations (see Churchman, 1979; Midgley, 1992, 2000; Curry and Pillay, 2009). However, this study nevertheless marks an advance in identifying influential agents in FRM in Curacao. Further research including a longitudinal study with a larger sample size could provide a basis for revealing patterns of power dynamics over time to predict or identify how changing power relations may impact on FRM regimes. Awareness of such changing dynamics can help plan initiatives that meet the interest of multiple parties, ensuring inclusive and effective FRM.

References

Agrawala, S., Carraro, M., Kingsmill, N., Lanzi, E., and Prudent-Richard, G. (2011). Private sector engagement in adaptation to climate change. OECD (Organisation for Economic Cooperation and Development) Environment Working Papers, No. 39. OECD, Paris.
Ailon, G. (2006). What B would otherwise do: A critique of conceptualizations of 'power' in organizational theory. *Organization*, 13(6), 771–800.
Allen, J. (2009). Three spaces of power: Territory, networks, plus a topological twist in the tale of domination and authority. *Journal of Power*, 2(2), 197–212.

Bachrach, P. and Baratz, M. S. (1962). Two faces of power. *American Political Science Review*, 56, 947–952.

Biagini, B. and Miller, A. (2013). Engaging the private sector in adaptation to climate change in developing countries: Importance, status, and challenges. *Climate and Development*, 5, 242–252.

Bryson, J. M. (2004). What to do when stakeholders matter: a guide to stakeholder identification and analysis technique. Paper presented at the London School of Economics and Political Science, London, 10 February.

Buzinde, C. N., Manuel-Navarrete, D., Yoo, E. and Morais, D. (2010). Tourists' perceptions in a climate of change: Eroding destinations. *Annals of Tourism Research*, 37(2), 333–354.

Callon, M. (1986). The sociology of an actor-network: The case of the electric vehicle. In Callon, M. (ed.), *Mapping the dynamics of science and technology*, chapter 2. Palgrave Macmillan, London.

Cameron, C. (2011). Climate change financing and aid effectiveness. Technical report, Agulhas Applied Knowledge, London.

Churchman, C. (1979). *The systems approach and its enemies*. Basic Books, New York.

Clayton, A. and Karagiannis, N. (2008). Sustainable tourism in the Caribbean: Alternative policy considerations. *Tourism and Hospitality Planning & Development* 5(3), 185–201.

Clegg, S. R. (1989). *Frameworks of power*. SAGE, London.

Curry, N. and Pillay, P. (2009). Converting food waste to usable energy in the urban environment through anaerobic digestion. 2009 IEEE Electrical Power and Energy Conference (EPEC), Montreal.

Dahl, R. A. (1957). The concept of power. *Behavioral Science*, 2(3), 201–215.

Dinica, V. (2006). Sustainable tourism development on Curacao: The implementation challenge. *WIT Transactions on Ecology and the Environment*, 97.

Dodman, D. and Mitlin, D. (2013). Challenges for community-based adaptation: Discovering the potential for transformation. *Journal of International Development*, 25(5), 640–659.

Dredge, D. and Jenkins, J. M. (2007). *Tourism planning and policy*. John Wiley & Sons Australia, Milton, Queensland.

Duval, D. (2004). *Tourism in the Caribbean: Trends, development, prospects*. Routledge, London and New York.

Eden, C. and Ackermann, F. (2013). Making strategy: The journey of strategic management. *International Journal of Market Research*, 43(2), 242.

Emerson, R. M. (1962). Power-dependence relations. *American Sociological Review*, 27, 31–41.

Granderson, A. A. (2014). Making sense of climate change risks and responses at the community level: A cultural-political lens. *Climate Risk Management*, 3, 55–64.

Hall, C. M. (2001). Trends in ocean and coastal tourism: The end of the last frontier? *Ocean and Coastal Management*, 44, 601–618.

Harris, G. (2010). Seeking sustainability in an age of complexity. *Journal of Environmental Management and Tourism*, 1(1), 63–78.

Honey, M. and Krantz, D. (2007). Global trends in coastal tourism. Technical report, Center on Ecotourism and Sustainable Development, Washington, DC.

Jennings, S. (2004). Coastal tourism and shoreline management. *Annuals of Tourism Research*, 31(4), 899–922.

Johnson, C. L., Tunstall, S. M. and Penning-Rowsell, E. C. (2005). Floods as catalysts for policy change: Historical lessons from England and Wales. *International Journal of Water Resources Development*, 21(4), 561–575.

Karagiannis, N. (2002). *Developmental policy and the state: The European Union, East Asia, and the Caribbean*. Lexington Books, Lanham, MD.

Kingdon, J. (1984). *Agendas, alternatives, and public policies*. Longman Higher Education, Boston.

Knoke, D. and Yang, S. (2008). *Social network analysis: Quantitative applications in the social sciences*. 2nd edition. Sage Publications, Thousand Oaks, CA.

Lukes, S. (2004). *Power: A radical view*. 2nd edition. MacMillan, London.

Matthews, D. (2008). Metadecision making: Rehabilitating interdisciplinarity in the decision sciences. *Systems Research and Behavioral Science*, 25(2), 157–179.

Mayers, J. and Vermeulen, S. (2005). Stakeholder influence mapping. Technical report, International Institute for Environment and Development (IIED), London.

McDavid, H. (2002). Why should government intervene in a market economy? A Caribbean perspective on the hospitality and tourism sector. In Boxill, I., Taylor, O., and Maerk, J. (eds), *Tourism and change in the Caribbean and Latin America*, 56–81. Arawak Publications, Kingston.

Meteorological Department Curacao (2015). Press release: World Meteorological Day. meteo.cw/Data_www/pdf/pub/Press Release_WMD2015.pdf (accessed 3 December 2017).

Meteorological Department Curacao (2017). Department mission. http://meteo.cw/about.php?Lang=Eng&St=TNCC&Sws=R11 (accessed 13 December 2017).

Midgley, G. (1992). The sacred and profane in critical systems thinking. *Systems Practice*, 5(1), 5–16.

Midgley, G. (2000). *Systemic intervention: Philosophy, methodology and practice*. Springer, New York.

Muis, S., Verlaan, M., Winsemius, H. C., Aerts, J. C. and Ward, P. J. (2016). A global reanalysis of storm surges and extreme sea levels. *Nature Communications*, 7, 11969–11980.

Mycoo, M. and Donovan, M. G. (2017). A blue urban agenda: Adapting to climate change in the coastal cities of Caribbean and Pacific small island developing states. Technical report, Inter-American Development Bank, New York.

NASA (2017). Shuttle Radar Topography Mission elevation data. www2.jpl.nasa.gov/srtm/ (accessed 21 November 2017).

Nurse, L., Sem, G., Hay, J., Suarez, A., Wong, P., Briguglio, S. and Ragoonaden, S. (2001). Small island states. In McCarthy, J. J., Canziani, O. F., Leary, N. A., Dokken, D. J. and White, K. S. (eds), *Climate Change 2001: Impacts, adaptation, and vulnerability. Contribution of Working Group II to the Third Assessment Report of the Intergovernmental Panel on Climate Change*, 842–975. Cambridge University Press, Cambridge.

Pauw, P. and Pegels, A. (2013). Private sector engagement in climate change adaptation in least developed countries: an exploration. *Climate and Development*, 5(4), 257–267.

Pelling, M. and Uitto, J. I. (2001). Small island developing states: Natural disaster vulnerability and global change. *Environmental Hazards*, 3(2), 49–62.

Pottier, N., Penning-Rowsell, E., Tunstall, S. and Hubert, G. (2005). Land use and flood protection: Contrasting approaches and outcomes in France and in England and Wales. *Applied Geography*, 25(1), 1–27.

Priestley, G. and Mundet, L. (1998). The post-stagnation phase of the resort cycle. *Annals of Tourism Research*, 25(1), 85–111.

Robinson, S. (2017). Climate change adaptation trends in small island developing states. *Mitigation and Adaptation Strategies for Global Change*, 22, 669–691.

Scott, D., Simpson, M. C. and Sim, R. (2012). The vulnerability of Caribbean coastal tourism to scenarios of climate change related sea level rise. *Journal of Sustainable Tourism*, 20(6), 883–898.

Self, J. and Penning-Rowsell, E. C. (2017). Power and policy in floodplain management, drawing on research in Alberta, Canada. *Water Policy*, 20(1), 21–36.

Simpson, M., Scott, D., Harrison, M., Silver, N., O'Keeffe, E., Harrison, S., Taylor, M., Sim, R., Lizcano, G., Wilson, M., Rutty, M., Stager, H., Oldham, J., New, M., Clarke, J., Day, O., Fields, N., Georges, J., Waithe, R. and McSharry, P. (2010). Modelling the transformational impacts and costs of sea level rise in the Caribbean. www.bb.undp.org/content/barbados/en/home/library/environment_energy/modelling-the-impacts-and-costs-of-slr-in-the-cbean.html (accessed 12 December 2017).

Sova, C. A., Helfgott, A., Chaudhury, A. S., Matthews, D., Thornton, T. and Vermeulen, S. (2015). Multi-level stakeholder influence mapping: Visualizing power relations across actor levels in Nepal's agricultural climate change adaptation regime. *Systemic Practice and Action Research*, 28(4), 383–409.

Sova, C. A., Thornton, T. F., Zougmore, R., Helfgott, A. and Chaudhury, A. S. (2017). Power and influence mapping in Ghana's agricultural adaptation policy regime. *Climate and Development*, 9(5), 399–414.

Tansey, O. (2007). Process tracing and elite interviewing: A case for non-probability sampling. *PS: Political Science & Politics*, 40(4), 765–772.

The Ministry of Health Environment and Nature (2014). National report of Curacao. www.sids2014.org/content/documents/599National%20Report%20Curacao%202014%20Sustainable%20Development%20FINAL%20%28July%202014%29.pdf (accessed 1 November 2017).

Tompkins, E. L. (2005). Planning for climate change in small islands: Insights from national hurricane preparedness in the Cayman Islands. *Global Environmental Change*, 15(2), 139–149.

Tompkins, E. L., Nicholson-Cole, S. A., Hurlston, L. A., Boyd, E., Hodge, G. B., Clarke, J., Gray, G., Trotz, N. and Varlack, L. (2005). Surviving climate change in small islands: A guidebook. https://inis.iaea.org/search/search.aspx?orig_q=RN:37003788 (accessed 15 December 2017).

Tunstall, S., McCarthy, S. and Faulkner, H. (2009). Flood risk management and planning policy in a time of policy transition. *Journal of Flood Risk Management*, 2(3), 159–169.

Weber, M. (1947). *The theory of social and economic organization.* Collier Macmillan, London.

Wilby, R. L. and Dessai, S. (2010). Robust adaptation to climate change. *Weather*, 65(7), 180–185.

Wilkinson, P. F. (2004). Tourism policy and planning: Case studies from the Commonwealth Caribbean. Cognizant Communication Corp, Elmsford.

5

POWER SHIFTS IN FLOOD RISK MANAGEMENT

Insights from Italy

Andrea Farcomeni

Introduction

Italy is particularly prone to floods and other natural hazards. Rarely do we witness years without flood episodes and victims in some parts of the country. The Italian Institute for Environmental Protection and Research (ISPRA) estimates that more than six million people are potentially exposed to floods (ISPRA, 2013), representing some 10 per cent of the Italian population.

This situation is the result of natural features such as geography, topography and climate, but also human behaviour and activities, which very often increase the degree of exposure and risk. Nevertheless, Italian legislation on this subject was neither coherent nor accurate until the first systematic laws were approved in 1989 and, more recently, in 2010. The inconsistency of flood risk management policy is surprising given that the death toll from flood events has been very high, counting 1,522 people during the period 1951–2012 (ISPRA, 2014).

This chapter evaluates power shifts that have occurred in flood risk management in Italy, with the main objective of understanding how institutional arrangements and negotiation between central and local forces have contributed to policy development in this sector. Following an integrative approach to negotiation, it shows that the shift towards decentralisation in this field must be read within the context of the entire restructuring of the Italian state and that the proposed devolution of power was merely instrumental to the interests of political elites who tried to obtain new support by changing the politics of scale.

Power and scalar processes

It is worth defining at the outset here what is meant by the word *power*, how power is conceptualised and its relation with the notion of scale. Having reflected on these concepts, we will then trace the dynamics of flood governance in Italy.

Power theory

Power is a highly contentious notion, whose meaning has been discussed by philosophers and thinkers such as Aristotle, Locke and Rousseau throughout the centuries. Dahl (1957, 203) depicted power as the ability of A to 'get B to do something that B would not otherwise do', but this assertion has since been contested by other authors (e.g. Bachrach and Baratz, 1962; Emerson, 1962; Lukes, 1974; Clegg, 1989). This chapter takes a pluralist approach to power – unlike elitists who argue that power is extremely centralised – and regards power as distributed throughout society (Bachrach and Baratz, 1962). From this viewpoint, the analysis stresses the dynamism embedded in the concept of power, its fluid distribution across different institutions and societal structures, and the fact that it is often spatialised in several territorial entities (for example, see Self and Penning-Rowsell, 2017). Starting from policy evolution and policy outcomes (Lukes, 1974), our investigation aims at identifying where power in flood risk management lies.

Nevertheless, we are not using the pluralist framework exclusively. Instead, what drives this investigation is the widespread significance given in the literature to the relational nature of power, which reveals itself in the complexity of the interactions between agents in the modern society, including individuals but especially institutions: 'Power is a property of the social relation' (Emerson, 1962, 32). The idea of power used in this chapter coincides with both the representation in critical decision-making processes and the actual capacity of a political actor to influence decisions regarding flood mitigation schemes and interventions, flood risk zoning and expenditure allocation. In this sense, it follows the first dimension or face of power (Lukes, 1974), which corresponds to the concrete and visible power to decide on actions and make decisions. On the other hand, it is also the power to determine the future evolution of the policy in this specific field.

Politics of scale

The conception of power as relative and subject to a relational dimension has repercussions on how we describe scalar processes and the politics of scale in a specific territory. The literature on the conception of scales has focused on emphasising its social constructive dimension as opposed to the previous naive interpretation that looks at scalar processes and territorial reconstructions as predetermined and permanent ontological entities (Delaney and Leitner, 1997; Swyngedouw, 1997; Cox, 1998; Brenner, 2001).

While we do not want to address the debate about the epistemological foundations of this contested socio-spatial concept, it is worth underlining that two approaches to the matter exist (MacKinnon, 2011): the political-economic approach and the post-structural approach, where the former sees scales as material entities while the latter considers them as theoretical construct without any onto-logical significance.

Bearing in mind this twofold partition, this chapter deals with the inherent relationship formed between the configuration of spatial scales, for instance the notion of 'national' and 'regional', and the distribution of power among actors applied specifically to flood risk management. Therefore, the goal is to apply that theoretical framework to flood risk management issues and hence describe where power emerges from the interactions of spatial relations and scalar reconstructions (Kaiser and Nikiforova, 2008; MacKinnon, 2011). To achieve a full understanding, the literature concerning devolution and decentralisation in Italy will act as central theme throughout the chapter.

Negotiation theory and its application to Italy

The theory that will help us appreciate the underlying process to be illuminated here concerns the negotiation and bargaining that has occurred between central and local authorities in Italy.

Negotiation theory is of paramount importance when dealing with decisions that involved mediations and conflicts between competing interests (Pardoe et al., 2011). Central to the analysis presented here is the role that negotiation has in setting agendas, engaging stakeholders and determining policy outcomes (Alfredson and Cungu, 2008). Several negotiation approaches have been elaborated to account for the nuances implicated during the bargaining procedure. Most of them stress the value of the notion of process, but emphasis is given to different peculiar aspects depending on the background discipline used as the theoretical framework.

Thus, the *structural approach* stresses the importance of the configuration of features bestowed to each negotiator, either an individual or an institution. Outcomes are perceived as deterministic and power-dependent results based on these characteristics (Bacharach and Lawler, 1981). Power asymmetry is the main factor through which one agent can dictate the terms of the agreement (Dawson, 1999). The *strategic approach* focuses on the function of goals in shaping outcomes and, drawing on game theory and rational choice theory, prescribing ways to act in order to achieve the best solution in a negotiation (Raiffa, 2002). The *behavioural approach* states that negotiators' individual characteristics determine the outcomes and the progress of the negotiation. The *processual approach* fundamentally sees negotiation as a process where negotiators display their positions and then try to converge, thus establishing agreeable compromises (Zartman, 1978). The *integrative approach* frames it as a sequence of phases in which the parties look for shared values to enable win-win solutions (Ury et al., 1991; Thompson, 2005).

These distinct but not mutually exclusive theories will be applied to our analysis to assess which one represents best the development of decision-making processes in flood risk management in Italy. We will use both the power and negotiation theories to analyse, below, the evolution of those processes. Table 5.1 summarises

TABLE 5.1 Negotiation and power distribution in different flood policy epochs in Italy

Flood policy epoch	Negotiation	Locus of power
Bargained centralism (1900–1922)	Win-win solutions: local elites trade central legitimacy for the protection of their entrenched interests and political status	Central state
Strong centralism (1922–1970)	Absence of negotiation; local forces are excluded from the political process	Central state
Decentralisation (1970–1988)	Regions claim and obtain more decision-making power	Regions
Process of layering (1989–2006)	Central government tries to take back control of lost powers: negotiation brings about shared responsibilities and pre-empts the resurgence of win-win solutions	Shared responsibilities
Bargained centralism (2010–to date)	Win-win solutions: climax of the layering process that creates an overarching central structure disguised as federalism	Central state (via Civil Protection Department)

where the locus of power lies in the key policy epochs examined in this chapter and the form of negotiation strategy that was used in each.

Flood risk management in Italy before the world wars: bargained centralism

The beginning of flood risk management in Italy can be traced back to the first years of the twentieth century with the Royal Decree 523 of 1904: the Consolidation Act on Water Works. At that time, the absence of expertise in the field urged the royal government to draw upon measures previously taken by pre-unification Italian kingdoms. The most prominent goal of this legislation was to set economic and institutional arrangements for land drainage, irrigation works and river engineering.

The Act divided hydraulic works into five categories, each of which was financially supported by different degrees of central government's commitment. Such projects were sustained by the need for maintaining agricultural production and the government's will to eradicate widespread malaria, which was still endemic in some parts of the country. There was no intention to promote a different approach based on mitigation measures and flood risk management. Indeed, the concept of prevention and risk-based interventions would only be adopted later, in 1989 (Testella, 2011).

Although a notion of cost sharing was acknowledged, the central government retained substantial decision-making power and economic control. The expenditure needed for works was so divided: 50 per cent by the central state; 30 per cent by Land Reclamation and Drainage Boards; 10 per cent by provinces; 10 per cent by local councils. These funding streams outline a situation where the central government had a strong control over decisions and capital allocation, balanced by interests and demands coming from local drainage boards and the farming lobby (cf. Tunstall et al., 2004). Accordingly, trade-offs between local and central interests were the expression of a distinguishing political dynamic typical of the Italian Liberal Age (1861–1913).

During this phase of Italian politics, the debate around the best form of government led to a symptomatic relationship between central authorities, represented by Piedmontese promoters of the unification, and more localised forces in the rest of Italy, especially in the south. The result was a 'bargained centralism' (Astuto, 2011; Baldini and Baldi, 2013), through which local elites traded central legitimacy for the protection of their entrenched interests and political status. Therefore, the constant bargaining and mediation between central and local forces gave birth to a centralist approach to flood risk management, characterised by hard and fiscal power of the central governments over provinces and local councils. Following an *integrative* interpretation to this negotiation, the two parties contending the power over decisions have collaborated to draft what eventually is a win–win agreement.

Strong centralism (1922–1970)

Following this first attempt to regulate and control watercourses to enhance agricultural productivity and to eliminate malaria, the most distinctive features of flood risk management in the following years were a persistent lack of enforcement, an absence of new legislation and static scales of governance.

The explanation can partly be given by the establishment of an authoritative government led by Benito Mussolini (1922–1943), who used his power to reinforce centralism and restrain local authority expenditures (Baldini and Baldi, 2013). During the fascist regime, the minister of the interior had the facility to appoint bodies that substituted municipal and provincial councils: 'a tight system of central controls was set up, and an overall policy of (political and financial) centralization was developed' (Baldini and Baldi, 2013, 96). This situation suddenly interrupted the negotiation between central government and local forces commenced soon after the unification.

After the end of the fascist era in 1945, attempts were made to include flood risk management within policy-making, but the new legislation was entirely biased towards landslide hazards. Consequently, Law 632 of 1967 allocated funding to protect mountainous areas from deforestation and soil erosion, whereas only limited funds were given to flood resilience. Nevertheless, besides funding commitment, there was no recommendation around means and interventions, while norms concerning floods were targeted at post-event interventions.

The lack of a specific set of measures for flood mitigation and management is surprising given the disasters that had occurred before the promulgation of Law 632: the flood in Polesine (an area 50 miles south of Venice) in 1951, the flood along the Amalfi Coast in 1954 and the widespread event in 1966 which flooded, amongst other places, the city of Florence. These three episodes accounted for 437 deaths. In fact, the flood in Florence urged the government to establish a committee (Commissione De Marchi) with the intent of handling the issue of flood risk management (Ministero dell'Ambiente, 2013). A window of opportunity was open (Kingdon, 1984) and the appointed committee worked for four years to produce a report issued in 1970.

The aim of the report was to establish a clear definition of 'soil conservation measures', a package of interventions to reduce natural hazards. The proposed means was urban spatial planning, through the concept of the 'Catchment Plan'. However, the suggestions from the committee did not act as a widespread catalyst for change, nor did they motivate the government to move towards a consistent risk-based approach. The notion of control over nature, exemplified by major structural engineering works, was still dominant and the capacity to exert power to comply with that principle was in the central state's hands. Although the republican Constitution promulgated in 1948 explicitly favoured a more decentralised state apparatus, the formal transfer of some powers to regions, including those related to flood risk management, would only happen after 1970.

The decentralisation process and the new role of regions (1970–2006)

Policies in flood risk management remained static in the period 1967–1989, while governments were trying to embrace the outcomes of the Commissione De Marchi in 1970. The ideas proposed by the committee were partly adopted a few years later in Law 171 of 1973, Measures for the Protection of Venice. This law was part of the wider Special Law of Venice, a set of specific rules around the protection of the Venetian Lagoon. This law is a precursor of the scalar reconstruction and power shift towards more decentralised stakeholders that the continuous tension between central government and regions were creating.

The Veneto region was given the power to elaborate a district plan for Venice and its surrounding towns, while a special commission was established to oversee the compliance of urban local plans with the district plan. At the same time, powers concerning flood prevention and hydraulic works remained a central state prerogative. Again, following an *integrative approach*, both parties were trying to achieve compromises based on shared values and eventually crafted win-win solutions.

Undoubtedly, the process of reconstruction was part of a wider debate that developed in Italy during the 1970s about the progressive decentralisation of competencies to regions (Roux, 2008; Keating and Wilson, 2010; Baldini and Baldi, 2013). For this reason, the shift towards decentralisation must be interpreted as part of the comprehensive reorganisation of the Italian state. Presidential Decree 616 of 1977 sanctioned a new scalar reorganisation by transferring several roles and functions to regions. Although the central government retained important financial powers, functions related to post-event flood interventions and protection measures were scaled down to the regions. Regions could decide what type of measures to adopt and their spatial dimension, but the central state tightly maintained decision-making powers regarding funding allocation. Furthermore, the central government withdrew financial resources from the National Solidarity Fund and distributed them to regions based on their need.

According to a 2013 Ministry of the Environment report, the reconfiguration created 'difficult and complex institutional arrangements owing to administrative uncertainties, fragmentation of governance systems, overlapping of competences, a power vacuum and sharp tensions between central and local authorities' (Ministero dell'Ambiente, 2013, author's translation). During this period and until 1989, the regions experienced the peak of their power, gaining legitimacy and political weight in their negotiation with the central state.

The process of scalar reconstruction culminated with Law 183 of 1989, Rules for the Reorganisation of Soil Conservation. It was an innovative piece of legislation through which the central government put forward the concept of 'catchment-wide planning'. It established basin authorities to rationalise and concentrate decision-making powers in one entity, introducing the notion of risk-based actions and mitigation.

Importantly, this Act influenced the course of scalar reconfiguration. Despite a gradual tendency towards decentralisation, the Law 183 exemplified a situation of conflict in which the central government was trying to regain some of the power eroded in 1977 while the regions were pushing for further reforms and autonomy. Consequently, the result was the creation of three types of basin authority: 11 national, 18 inter-regional and an undefined number of regional basin authorities. The technical committees of these authorities were chaired by both central state and regional officials. Overall, this new Act sought to introduce a new integrated approach between spatial planning, water resources and land use. The principal means to achieve this ambitious goal was the Hydrogeological Setting Plan (Piano di Assetto Idrogeologico), part of the more comprehensive 'Basin Plan'.

The scalar arrangement of the Hydrogeological Setting Plan introduced in 1989 lasted for 17 years until 2006. During this period, several flood events hit the country, the most devastating being the Sarno flood and landslide in 1998, which killed 159 people in the province of Salerno (ISPRA, 2013). The status quo was modified once again in 2006. Legislative Decree 152 of 2006, known as the Environmental Code, abrogated Law 183, Rules for the Reorganisation of Soil Conservation, and reshaped the governance system. The normative structure remained roughly the same, except for the introduction of three-year programmes to monitor the progress of interventions. The basin authorities were abolished and eight district basin authorities were created instead.

Furthermore, this new piece of legislation established a Public Agencies Conference, a stakeholder forum to facilitate the dialogue between central authorities and regions. The Conference, however, stated non-binding opinions, while final decisions and financial powers rested with the central government. This last piece of legislation clearly shows the attempt of the central state to balance some of the power that the regions gained during the years of decentralisation in the 1970s and 1980s. It highlights the underlying negotiation that the new governance structure imposed, during which the convergence of central and regional interests pre-empted the resurgence of win-win solutions in the years that followed.

Back to bargained centralism and the new role of Civil Protection

In recent years, the distribution of power has undergone a further reorganisation following the emergence of new dynamics at both national and international level.

First, the influence of the European Union and its 2007 Floods Directive (2007/60/EC) has encouraged the Italian government to fill the gaps that the previous legislation had left. Second, the inconsistency of enforcement and the persistence of ad hoc interventions (Testella, 2011) have allowed the Civil Protection Department to become a powerful new player. This department is a national body that deals with the prevention, forecast and monitoring of risks as well as the coordination of natural disaster response. The new arrangement is clearly reflected in the latest law on flood risk management, Legislative Decree 49 of 2010, which acknowledges the European Floods Directive 2007/60/EC. This decree defines roles and responsibilities related to the formulation of the Flood Risk Management Plan.

Within this framework, district basin authorities oversee preparing flood hazard maps while regions, together with the Civil Protection Department, draft the part of the plan concerning flood warning systems. It can be clearly observed how the movement towards a more decentralised configuration has been lessened by a strong representation of the central government during both the stages of the Plan. In effect, the district basin authorities are governmental bodies chaired by both central state officials and region delegates, while the Civil Protection Department is under direct control of the president of the Council of Ministers. The overall architecture is far from the decentralisation appearing during the 1970s and 1980s, and the central grip has remained strong principally because what emerges is a perpetual negotiation between forces and their ability to systematically converge to obtain shared compromises.

Therefore, the emergence of the Civil Protection Department as an important player in the flood risk management discourse belongs to this political game. Its role is not confined solely to alert systems but embraces a vast spectrum of tasks, comprising the management of emergencies after natural disasters, the coordination of reliefs and post-event measures (Klijn et al., 2008). Hence, when the central government declares a state of emergency following flood events, the Civil Protection Department, through the appointment of a commissariat, takes over all the responsibilities concerning land governance and issues ordinances that replace ordinary administration acts (Alexander, 2002). Since in many cases the declared state of emergency continues after the real emergency is over, the Civil Protection Department has the power to manage and influence spatial planning and flood risk management in areas hit by flooding.

The result is the partial dismissal of the previous legislation on flood risk management and less ability from decentralised bodies to have a say in decisions that affect local spatial planning and flood resilience. Within this context, decision-making forums where central government and regions coexist are by-passed in favour of a centralised body that can discipline flood risk management and manage funding allocations. At the same time, local and regional political interests are preserved in exchange for central legitimacy and a prestigious but ineffective participation in decision-making bodies. This aspect reinforces the idea that flood risk management in Italy has broadly taken an *integrative approach* to negotiation.

Conclusion: centralism in disguise

We have shown that power in flood risk management in Italy has been characterised by a process of scalar reconfiguration that has not fundamentally changed the hidden drivers that influence power distribution.

At the beginning of the twentieth century, the central government and land drainage authorities controlled major structural works aimed at defending agricultural productivity. Progressively, regions undermined that central authority control, which, in turn, reacted by imposing shared responsibilities in the decision-making process. The result is an emergency-based framework, within which the Civil Protection Department is now polarising part of the power. By taking a pluralist approach we have described the dynamism embedded in the concept of power and its fluid distribution across different institutions and societal structures. Central to the analysis is the role that negotiation has in setting agendas, engaging stakeholders and determining flood risk management policy outcomes.

This chapter demonstrates that this negotiation in Italy follows an *integrative* interpretation, where the dynamics are the result of the constant mediation between central and local forces which have drafted what is eventually a win–win agreement, cemented in an anomalous but profitable alliance. Although regions have lost much of the power they gained during the 1970s and 1980s, the central-local alliance continues to exist, perpetuating the strong centralist attitude entrenched in Italian political culture.

Importantly, the pattern of change in flood risk management closely follows those in the surrounding political background. The shift towards decentralisation in this field must be read through the analytic lens of the entire restructuring of the Italian state. As with other dimensions of policy-making, the proposed devolution was merely instrumental to the interests of political elites who tried to obtain new support by changing the politics of scale and adapting the old apparatus to the new needs (Baldini and Baldi, 2013). The method used was what Baldini and Baldi (2013) called a process of 'layering'. By simply adding new institutional frameworks, the central government has managed to create an overarching central structure disguised as genuine federalism. The result has been a complexity of bodies, responsibilities and jurisdictions that allow the central government to retain much of its power in decision-making procedures through its coexistence in key decision bodies.

References

Alexander, D. (2002). The evolution of civil protection in modern Italy. In Dickie, J. and Foot, J. (eds) *Disastro! Disasters in Italy since 1860: Culture, politics, society*. Palgrave, New York, 165–185.

Alfredson, T. and Cungu, A. (2008). Negotiation theory and practice: A review of the literature. FAO Policy Learning Programme, EASYPol Module 179.

Astuto, G. (2011). *L'amministrazione italiana. Dal centralismo napoleonico al federalismo amministrativo*. Carrocci, Rome.

Bacharach, S. and Lawler, E. (1981). Power and tactics in bargaining. *Industrial and Labour Relations Revie*, 34(2), 219–233.

Bachrach, P. and Baratz, M. S. (1962). Two faces of power. *The American Political Science Review*, 56(4), 947–952.

BaldiniG. and BaldiB. (2013). Decentralization in Italy and the trouble of federalization. *Regional and Federal Studies*. Published online: 11 Sept.

Brenner, N. (2001). The limits to scale? Methodological reflections on scalar structuration. *Progress in Human Geography*, 25(4), 591–614.

Clegg, S. (1989). *Frameworks of power*. Sage Publications, London.

Cox, K. (1998). Spaces of dependence, spaces of engagement and the politics of scale, or: looking for local politics. *Political Geography*, 17, 1–23.

Dahl, R. (1957). The concept of power. *Behavioural Science*, 2(3), 201–215.

Dawson, R. (1999). *Secrets of power negotiating: Inside secrets from a master negotiator*. Career Press, London.

Delaney, D. and Leitner, H. (1997). The political construction of scale. *Political Geography*, 16, 93–97.

Emerson, R. M. (1962). Power-dependence relations. *American Sociological Review*, 27(1), 31–41.

ISPRA (Istituto Superiore per la Protezione e la Ricerca Ambientale) (2013). Tematiche in primo piano: Annuario dei dati ambientali 2012. Stato dell'ambiente 39/2013.

ISPRA (Istituto Superiore per la Protezione e la Ricerca Ambientale) (2014). Annuario dati ambientali. http://annuario.isprambiente.it/content/versioni. Last access: 17 January 2014.

Kaiser, R. and Nikiforova, E. (2008). The performativity of scale: the social construction of scale effects in Narva, Estonia. *Environment and Planning D*, 26(3), 537–562.

Keating, M. and Wilson, A. (2010). Federalism and decentralisation in Italy. Political Studies Association (PSA) Conference, Edinburgh, March–April 2010.

Kingdon, J.W. (1984). *Agendas, alternatives, and public policies*. Little, Brown, Boston.

Klijn, F., Samuels, P. and Van Os, A. (2008). Towards flood risk management in the EU: state of affairs with examples from various European countries. *International Journal of River Basin Management*, 6(4), 307–321.

LukesS. (1974). *Power: A radical view*. Macmillan, New York.

MacKinnon, D. (2011). Reconstructing scale: Towards a new scalar politics. *Progress in Human Geography*, 35(1), 21–36.

Ministero dell'Ambiente (2013). Documento conclusivo del tavolo tecnico Stato-Regioni, Indirizzi operativi per l'attuazione della Direttiva 2007/60/CE relativa alla valutazione ed alla gestione dei rischi da alluvioni con riferimento alla predisposizione delle mappe della pericolosità e del rischio di alluvioni. Decreto Legislativo n. 49/2010.

Pardoe, J., Penning-Rowsell, E. C. and Tunstall, S. (2011). Floodplain conflicts: Regulation and negotiation. *Natural Hazards and Earth System Science*, 11, 2889–2902.

Raiffa, H. (2002). *Negotiation analysis*. Belknap Press of Harvard University Press, Cambridge and London.

Roux, C. (2008). Italy's path to federalism: Origins and paradoxes. *Journal of Modern Italian Studies*, 13(3), 325–339.

Self, J. and Penning-Rowsell, E. C. (2017). Power and policy in floodplain management, drawing on research in Alberta, Canada. *Water Policy*, 20(1), 21–36.

SwyngedouwE., (1997). Excluding the other: The production of scale and scaled politics. In Lee, R. and Wills, J. (eds), *Geographies of economics*. Arnold, London, 167–176.

Testella, F. (2011). Evoluzione normative sul rischio idrogeologico dalla Legge 183/1989 alla Direttiva Alluvioni (2007/60/CE) e il Decreto Legislativo 49/2010. Slide presentation. CMCC (Centro Euro-Mediterraneo sui Cambiamenti Climatici), Venice, 12 September.

Thompson, L. L. (2005). *The mind and heart of the negotiator*. Prentice Hall, London.

Tunstall, S. M., Johnson, C. L., Penning-Rowsell, E. C. (2004). Flood hazard management in England and Wales: From land drainage to flood risk management. Proceedings of the World Congress on Natural Disaster Mitigation, 18–21 February, New Delhi, India.

Ury, W., Fisher, R. and Patton, B. (1991). *Getting to yes: Negotiating agreement without giving in*. Random House, New York.

ZartmanW. I. (1978). *The negotiation process: Theories and applications*. Sage Publications, Beverly Hills, CA.

6

'GOING DUTCH' IN FLOOD RISK MANAGEMENT

How is Dutch flood policy mobilised?

Timo Maas

Introduction

Centuries of struggling with water has given the Dutch water sector a worldwide reputation for knowing how to keep your feet dry (NWP, 2011). As a result, their expertise is often asked for elsewhere, illustrated for instance by the sector's immediate involvement following hurricanes Katrina and Sandy in the United States, or a call for a 'Dutch-style' flood plan following the December 2015 floods in the United Kingdom (van Klaveren, 2015).

In a world where non-state actors are gaining increasing influence in flood management (Meijerink and Dicke, 2008), this raises the question what 'going Dutch' means for (foreign) flood policy. The question is especially poignant given the increased attention to the value of local knowledge in flood policy (Whatmore, 2009) and the suggestion by Sultana et al. (2008) that non-state actors from foreign countries may have particularly significant influence in developing countries.

With this in mind, this chapter seeks to provide a better understanding of the way the practices and discourses of organisations in the Dutch water sector affect flood policy in these places. We do not attempt to provide answers as to the *results* of this influence, but rather to analyse how this *process* of transfer works in the case of flood risk management and 'the Dutch'. Accordingly, we follow a 'policy mobilities' approach, which directs attention to the 'actors, practices, and representations that affect the (re)production, adoption and travel of policies, and the best practice models across space and time' (Temenos and McCann, 2013, 345). Through this frame, we can obtain a nuanced and detailed view of what it means for a policy to be mobile and what the roles of actors involved in this process might be.

The chapter is structured as follows. The next part sets out the theoretical framework employed, discussing how the policy mobilities approach has been developed by combining elements from political science, sociology and geography.

The third part presents a history of Dutch flood policy and discusses the results of seven interviews with experts working in the Dutch water sector. The interviewees reflected on 'doing' flood risk management in developing countries, in a very broad way, and we highlight and discuss the implications of three aspects that seem especially interesting: formal institutions, local knowledge and the importance of the Dutch sector itself. We conclude in the fourth part with some final reflections.

Policy mobilities

From a theoretical perspective, policy mobilities are about the ways that elements of policies travel and who is involved in this travelling. The approach was developed by political geographers critiquing work in political science on 'policy transfer', introducing elements from sociological work on mobilities and geographical work on scale to address their critique (Temenos and McCann, 2013). Policy transfer studies the ways in which knowledge of the policies and institutions of one place is used to develop policy elsewhere (Benson and Jordan, 2011). It provides an important foundation to policy mobilities, but has been critiqued on three main aspects: (1) its predominantly nation-based approach; (2) its focus on rational, optimising agents, neglecting agency; and (3) a lack of attention to the changes that occur in policy on the move (McCann and Ward, 2012). A more comprehensive genealogy of the concept can be found in Temenos and McCann (2013), and for an overview of the debate between policy transfer and policy mobilities see, for example, Marsh and Evans (2012) and McCann and Ward (2013).

Seeing policies as *assemblages* provides a more dynamic understanding of their transfer. This means policies are not just a law or regulation, but a collection of related actors, texts and formal and informal institutions (Prince, 2010). In this, expertise and regulation come together. The observation that technical expertise and politics are two sides of the same coin is an important conclusion from the field of science and technology studies, which holds that 'the ways in which we know and represent the world (both nature and society) are inseparable from the ways in which we choose to live in it' (Jasanoff, 2004, 13)(see also chapter 15, herein).

For flood policy, this means that the way we conceptualise and analyse flood risk is crucial for understanding the way we attempt to address it. This also highlights the significance of the role of consultants doing flood risk calculations, making it important to think about the (implicit) assumptions and underlying discourses that they draw on (Larner and Laurie, 2010; Bevir 2011). See chapter 1, Information and knowledge, for more on this topic. Furthermore, this category of actors rarely has any formal power to influence policy (Stone, 2010). While Dutch water experts in developing countries have little agency when it comes to the formal processes of policy-making, informally their influence may be significant.

An important implication from this assemblage position for policy mobilities is that it means that a policy moving in space will *transform*. Not all elements of the 'original' policy will move with it, and work needs to be done to integrate the mobilised elements of the policy in the new context (Peck, 2011). Cook and Ward (2011) discuss this in their study of 'policy tourism' for large sporting events. The policy mobility that takes place is not simply a single copy-and-paste from one place to another, but is a learning process that happens through a multitude of points of comparison.

Attention to the mobility of people and technologies further directs attention to how policies are transferred in hierarchical and uneven ways (McCann, 2011); indeed, the simple observation that the Dutch water sector has the resources to transport its people around the world, attend conferences, subscribe to scholarly literatures, etc., is important to realise when discussing the sector's status and its influence in policies elsewhere. A mobilities perspective makes us aware of the circulation of policy knowledge, importantly the fact that it must somehow 'touch down', and emphasises how various levels of government interact in how this knowledge circulates (Temenos and McCann, 2013). Here, the approach builds on a geographical perspective of scale as socially produced and relational. For example, we can think of the increasing importance of cities in governing adaptation to climate change, while their policies are inextricably linked to what happens at other levels and circulate through several city networks (Bulkeley and Castán Broto, 2013).

A good example of how the policy mobilities frame can be operationalised is found in Wood's (2015) study of the adoption of bus rapid transit systems in South Africa. This transport system has become synonymous with Bogotá – although in that city too it has been the result of decades of learning within South America – and its improvements in economic and social wellbeing in the past two decades. Wood describes not only the groups advocating bus rapid transit to be used around the world, but also how South African policy makers believed that the context in Bogotá was similar to that in cities in South Africa.

In this process, Wood suggests they ignored the bus rapid transit system that already existed in Lagos, Nigeria, for political reasons: they believed South African cities were 'further advanced' than elsewhere in Africa. The result was that the South African bus rapid transport systems arguably focused on replicating the (technical) requirements of the 'best practice' found in Bogotá at the expense of fitting the system into the local transport context. Similarly, we can ask the question in what ways the Dutch water sector will employ their knowledge, developed in a particular context (The Netherlands), in their international activities.

How do the Dutch assemble flood policy elsewhere?

To gain better insight in the practices of Dutch water experts and the (implicit) assumptions that influence these, we held semi-structured interviews with seven experts, based at an applied research institute, consultancy and umbrella organisation (Table 6.1). Most of the interviewees have extensive experience with flood risk management projects in developing countries.

The interviews focused on the project cycle (from acquisition to dissemination), working with local partners and the way Dutch policy related to projects. It should

TABLE 6.1 Interviewee organisations

Interviewee	Organisation
1	Applied research institute
2	Consultancy
3	Applied research institute
4	Applied research institute
5	Umbrella organisation
6	Applied research institute
7	Applied research institute

be noted that the interviewees largely represent just one type of organisation, namely a public knowledge institute providing scientific input to (flood) policy-making (Koens et al., 2016). Interviewees indicated that projects try to find middle ground between innovation and practice, but at the same time this is also a process that is highly tied to the individual background and ambitions of individual practitioners (interviewee 7). From the myriad examples and nuances the interviews yielded, we will limit the discussion in this chapter to three aspects: formalised institutions, local knowledge and the Dutch presence itself. Before discussing these aspects, we provide a short history of Dutch flood risk management.

The origins of Dutch flood policy can be found in the establishment of regional water organisations called 'water boards' from around the thirteenth century onwards, as cooperation on the topic of flood defences and land drainage was recognised to be essential for efficiency gains (VanKoningsveld et al., 2008). Such cooperation between a wide range of individuals and institutions seeking consensus has later become known as the 'poldermodel'. From the seventeenth century onwards a range of 'megaworks' were implemented, first to drain inland lakes and later, in response to major coastal flooding in 1916 and 1953, to reduce significantly the length of the coastline and provide better protection from the sea.

More recently, 'near misses' of riverine flooding in 1993 and 1995 have led to the adoption of a more holistic approach (VanKoningsveld et al., 2008). This approach, called 'Room for the River', is a set of 39 measures implemented around 2015/16 to accommodate and work with flooding where possible, for instance through the widening and lowering of floodplains and providing areas for the temporary storage of water (Rijke et al., 2012). It is a unique programme through its emphasis on the 'poldermodel', with many measures being the result of extensive negotiations between actors as well as an interesting example of working with measures that aim to build 'resilience' (interviewee 4). As of 2010, the 'Deltaprogramme' forms the main Dutch policy for flood safety and freshwater supply, providing a further step into integrated water resources management.

Still, the long experience with flood defences in The Netherlands favours a conceptualisation of flood risk as 'probability times consequences', which favours the engineers' thinking in terms of flood defences (Klijn et al., 2015).

Internationally, both the high flood protection standards adopted as well as the shifting approach of the past decade have received considerable attention, touting both the centuries of experience as well as the innovative solutions of the 'Room for the River' programme (e.g. Katz, 2013; Kimmelman, 2013; Carrington, 2014; Horne, 2017).

As a result of the long history and recent developments in Dutch flood policy, the institutional landscape in relation to water management in The Netherlands is highly formalised. However, as our interviewees indicated, this is often not the case in developing countries, relating to at least three issues.

First, it is often not clear which organisation – including different levels of government – has a certain mandate, and overlap between organisations is a common occurrence. This can be especially problematic when a measure has been implemented, but it is unclear where the mandate (and thus also funding source) for operations and maintenance lies (interviewee 7).

A second issue is lack of cooperation between local institutions. In this context, interviewee 3 highlighted the importance of improving cooperation within the triangle of government, business, and knowledge institutes (or diamond, when NGOs are added as a fourth actor). Also within these categories of actors, cooperation can be improved, for instance by having multiple knowledge institutes cooperate. Currently a programme is underway in which Dutch and Indonesian knowledge institutes for water, meteorology and ecology join forces for more holistic planning on these domains (interviewee 6).

This is related to a third issue, namely that of data. Most of the work done by the Dutch water sector relies heavily on computer models and scenario analysis. These require a substantial amount of data, for which local partners are indicated by interviewees to be the primary source. However, in certain cases there are no formalised institutions that produce the long-term scenarios for local climate, population and other socio-economic factors that are required as input to these integrated models. Establishing these then becomes a prerequisite for detailed analyses of flood risk and potential interventions (interviewee 3). Chapter 10 explores this issue in further detail in a wider context.

This exemplifies the emphasis the policy mobilities framework puts on the various component parts of a policy. The Dutch water sector is accustomed to a specific approach to collecting the 'evidence' on which to base (further) flood policy, e.g. what measures can be implemented that would also be (cost-)effective. In this regard, cooperation with local partners is sought to obtain access to data that *does* exist, but also gain better insight into the institutional context of the area they are working in. This results in 'hybrid knowledge' obtained by combining external and local knowledge (Haughton et al., 2015). Part of this knowledge, introduced by locals, is what the current institutional context is. External experts subsequently promote a process of (further) institution and capacity building, based on the arrangements they know from elsewhere.

Our point here is not so much whether the (intended) outcomes of the subsequently implemented policy equal those in The Netherlands, but to illustrate how flood policy is largely enacted by 'middling technocrats' rather than by people in high status roles (Larner and Laurie, 2010). By seeing policy as an assemblage made up of many different elements, the policy mobilities framework turns our attention to the fact that the practices and models these 'middling technocrats' promote always make (implicit) assumptions about what is valuable and what is not. This means that in their everyday work, these consultants do not just mobilise their knowledge and approach to flood risk, but also the ethics they embed. These may for instance include the roles that various actors involved with flood risk have, from various level of government to businesses to individual citizens.

At the same time, our interviewees emphasised the need to work with the actual needs and desires of locals. Contention among interviewees existed regarding the extent to which this happens, relating closely to the way floods are managed in The Netherlands itself. Interviewee 6 (research institute) reflected on this contention, saying that 'our way of managing floods, including the norms and values it is based on, is not necessarily theirs', but also indicated that structural measures (dykes and other engineered solutions) will always be necessary. Another interviewee saw the clearest 'Dutch' influence exactly there, namely that 'when Dutch experts see an area that floods, they want to embank it' (interviewee 7). He related this to the way vulnerability is seen. In many flood models, land use is used in combination with inundation levels to compute a monetised damage value (Ward et al., 2013). As the absolute damage to, for instance, industrial areas is often higher than to agriculture, such land use changes can give good economic reason to embank a certain area over another. However, when considering resilience as the 'ability of the system to recover from floods' (de Bruijn, 2005, 2) it may be more appropriate to consider the *relative* damage done by a flood, as a subsistence farmer may struggle more to recover from a flood than an industrial area (interviewee 7). See chapter 13 for further discussion on social justice issues surrounding flood protection.

Synthesising from this point stresses that the long history of intervention in The Netherlands has made the country skilled at ensuring that water simply does not cross into undesired places (or only with very low annual probabilities, generally at return periods of 1 in 1,250 years or more). However, when priorities need to be set for managing flood *risk* instead, as is the case in developing countries, it is essential to make sure those affected have a strong voice. While a discussion of stakeholder participation is beyond the scope of this chapter, it is important to point out that there may be micro-political barriers to stakeholder engagement that limit the amount and quality of participatory processes (Tseng and Penning-Rowsell, 2012).

Furthermore, even if practitioners acknowledge and seek to include local knowledge in their work, (some) scientists and governments may still take a more technocratic view (Mercer et al., 2012). From the policy mobilities perspective, this means that the background and ambitions of individual practitioners are key elements of the way in which policy mutates while on the move. The way practitioners are educated is likely to be important here. Interdisciplinary skills to reflexively translate between governance, engineering and local cultures are necessary to do the work to re-adapt policies made mobile. Such skills thus facilitate the type of mutations necessary for context-sensitive policy mobilities.

Finally, it is important to reflect on the claim made in the introduction of this chapter, namely that the Dutch water sector is expected to be an important player in the international arena. This element is a good example of why the policy mobilities literature holds 'small p' politics above an 'optimizing, rational' search for policies to import (McCann and Ward, 2013). As mentioned above, international media frequently puts Dutch flood policy forward as exemplary. Dutch water experts have been said to hold 'almost mythical status' (Storr, 2014), and 'the world is watching' the solutions held by the Dutch (Kimmelman, 2017). The Netherlands is frequently visited by journalists and policy-makers (van Klaveren, 2015), a typical example of the 'policy tourism' through which officials learn about the policies of a particular place (Cook and Ward, 2011).

The sector's importance is further illustrated by the fact that Dutch organisations were involved in four or five out of ten consortia in the post-Hurricane Sandy efforts in New Jersey and New York (interviewee 5), as well as their work in New Orleans and in Colombia following major flood disasters (interviewee 2). Involving Dutch experts can thus be considered something that brings a certain legitimacy to flood policy making (Stone, 2004). In addition, our interviewees indicated that part of the reason why the Dutch are involved is to some degree the result of path dependency. Prior work makes it likely that certain models and other tools are available (interviewee 1), as well as that contacts with (potential) local partners exist (interviewee 5).

But like Wood's (2015) bus rapid transit example, a policy mobilities frame highlights that it is not just about the 'small p' politics on the 'receiving' side, but also about how policy models are promoted. Just like several advocacy groups promote bus rapid transit, there is an active effort made by the Dutch government and water sector to bring Dutch expertise to the market and to bring overseas students to Dutch Universities to study on water courses and research water issues in their home countries: this is not left to chance. The Netherlands Water Partnership (NWP) is an organisation aimed at the international market; it advocates joint branding of the Dutch water sector, which is achieved through seeking connections with 'traditional' Dutch branding (e.g. clogs and tulips) as well as through diplomatic channels (NWP, 2011). This ambition was further boosted with the appointment of a Special Envoy for International Water Affairs by the Dutch government in March 2015. NWP may also function as a platform that is useful to expand to new markets as well as to network *within* the Dutch water sector (interviewee 2).

Furthermore, multiple other such umbrella organisations analyse calls for tender offers by development banks and advise Dutch organisations on suitable consortia to respond to these tenders (interviewee 5). For instance, presentations of 'the Dutch approach' strengthen relations that result in early signals of approaching tenders, and can be used to actively market the approach itself (e.g. Beckers, 2013). To interviewee 5, such activities also assist these umbrella organisations in being a 'scout', as networking opportunities are created that in turn can lead to (informal) notifications of upcoming tender calls, allowing for more time to create a strong bid. These aspects, that are meant to actively bring Dutch policy knowledge in circulation, should be seen in conjunction with the possibilities for this knowledge to 'touch down' described in the previous paragraph. While considerable effort is put into making sure the Dutch water sector is 'ready to move', this also requires actors elsewhere to see it as important and thus be willing to receive it.

Conclusions

This chapter has explored what the influence of the Dutch water sector is on flood risk policy in developing countries. Through the 'policy mobilities' framework we have focused on the ways in which Dutch flood policy is mobilised and becomes a component part of the policies the sector helps install elsewhere. We have discussed three important aspects in this regard.

First, much of the work performed by Dutch water professionals is related to institution and capacity building, including creating (informal) links between local institutions. In this regard, the policy mobilities approach highlights the fact that particular knowledges and discourses underlying this work also embed particular ethics. This feeds in to the second aspect, namely that through valuing and incorporating local knowledge the 'Dutch predispositions' regarding flood risk management can be overcome. This requires individual professionals able to work reflexively, enabling context-sensitive policy mutations. The third aspect concerns the way the prominence of the Dutch sector is maintained and promoted. Here, the combination of international interest and Dutch efforts to arouse and maintain such interest keep Dutch policy knowledge mobile.

An important note at this point is that the Dutch involvement is by no means necessarily a bad thing. On the contrary, real problems with flooding exist in large parts of the world, and Dutch expertise can be of real value in reducing these problems. Nonetheless, it remains important for the Dutch water sector to be aware of both the 'privilege' that comes with its status, as well as keeping aware of the fact that Dutch water policy is done a certain way. As our interviewees were well aware: this is by no means a uniquely optimal way.

Acknowledgement

I am grateful to the interviewees for lending me their time and providing me with valuable insights; obviously, with regard to these, the usual disclaimer applies.

References

Beckers, J. 2013. Dutch approach to coastal flood hazard. NRC Workshop on Probabilistic Flood Hazard Assessment (PFHA), Rockville, MD, 29–31 January.
Benson, D. and Jordan, A. 2011. What have we learned from policy transfer research? Dolowitz and Marsh revisited. *Political Studies Review*, 9(3), 366–378.

Bevir, M. 2011. Governance and governmentality after neoliberalism. *Policy & Politics*, 39(4), 457–471.

de Bruijn, K. M. 2005. *Resilience and flood risk management: A systems approach applied to low-land rivers*. Delft: Delft University Press.

Bulkeley, H. and Castán Broto, V. 2013. Government by experiment? Global cities and the governing of climate change. *Transactions of the Institute of British Geographers*, 38(3), 361–375.

Carrington, D. 2014. Taming the floods, Dutch-style. *The Guardian*, 19 May.

Cook, I. R. and Ward, K. 2011. Trans-urban networks of learning, mega events and policy tourism. *Urban Studies*, 48(12), 2519–2535.

Haughton, G., Bankoff, G. and Coulthard, T. J. 2015. In search of 'lost' knowledge and outsourced expertise in flood risk management. *Transactions of the Institute of British Geographers*, 40(3), 375–386.

Horne, J. 2017. The Dutch understand flooding: Why can't America manage it? [online]. *CityLab*. Available from: www.citylab.com/environment/2017/08/flood-prevention-katrina-new-orleans-houston-harvey/538185/. [Accessed 28 July 18]

Jasanoff, S. 2004. *States of knowledge: The co-production of science and social order*. London: Routledge.

Katz, C. 2013. To control floods, the Dutch turn to nature for inspiration [online]. *Yale Environment 360*. Available from: https://e360.yale.edu/features/to_control_floods_the_dutch_turn_to_nature_for_inspiration. [Accessed 28 July 18]

Kimmelman, M. 2013. Going with the flow. *New York Times*, 13 Feb.

Kimmelman, M. 2017. The Dutch have solutions to rising seas: The world is watching. *New York Times*, 15 June.

van Klaveren, H. 2015. We need a Dutch-style delta plan to stem the tide of floods. *The Guardian*, 27 Dec.

Klijn, F., Kreibich, H., de Moel, H. and Penning-Rowsell, E. 2015. Adaptive flood risk management planning based on a comprehensive flood risk conceptualisation. *Mitigation and Adaptation Strategies for Global Change*, 20(6), 845–864.

Koens, L., Chiong Meza, C., Faasse, P. and de Jonge, J. 2016. Public knowledge organisations in the Netherlands. Rathenau Instituut, The Hague.

Larner, W. and Laurie, N. 2010. Travelling technocrats, embodied knowledges: Globalising privatisation in telecoms and water. *Geoforum*, 41(2), 218–226.

Marsh, D. and Evans, M. 2012. Policy transfer: Coming of age and learning from the experience. *Policy Studies*, 33(6), 477–481.

McCann, E. 2011. Urban policy mobilities and global circuits of knowledge: Toward a research agenda. *Annals of the Association of American Geographers*, 101(1), 107–130.

McCann, E. and Ward, K. 2012. Policy assemblages, mobilities and mutations: Toward a multidisciplinary conversation. *Political Studies Review*, 10(3), 325–332.

McCann, E. and Ward, K. 2013. A multi-disciplinary approach to policy transfer research: geographies, assemblages, mobilities and mutations. *Policy Studies*, 34(1), 2–18.

Meijerink, S. and Dicke, W. 2008. Shifts in the public–private divide in flood management. *International Journal of Water Resources Development*, 24(4), 499–512.

Mercer, J., Gaillard, J. C., Crowley, K., Shannon, R., Alexander, B., Day, S. and Becker, J. 2012. Culture and disaster risk reduction: Lessons and opportunities. *Environmental Hazards*, 11(2), 74–95.

NWP, 2011. Water 2020: Wereldleiders in water – De toekomstvisie van de Nederlandse watersector. NWP, The Hague.

Peck, J., 2011. Geographies of policy: From transfer-diffusion to mobility-mutation. *Progress in Human Geography*, 35(6), 773–797.

Prince, R., 2010. Policy transfer as policy assemblage: Making policy for the creative industries in New Zealand. *Environment and Planning A*, 42(1), 169–186.

Rijke, J., van Herk, S., Zevenbergen, C. and Ashley, R. 2012. Room for the River: delivering integrated river basin management in the Netherlands. *International Journal of River Basin Management*, 10(4), 369–382.

Stone, D., 2004. Transfer agents and global networks in the 'transnationalization' of policy. *Journal of European Public Policy*, 11(3), 545–566.

Stone, D., 2010. Private philanthropy or policy transfer? The transnational norms of the Open Society Institute. *Policy & Politics*, 38(2), 269–287.

Storr, W., 2014. How can Holland help us fight the floods? *The Telegraph*, 22 April.

Sultana, P., Johnson, C. and Thompson, P. 2008. The impact of major floods on flood risk policy evolution: Insights from Bangladesh. *International Journal of River Basin Management*, 6(4), 339–348.

Temenos, C. and McCann, E. 2013. Geographies of policy mobilities. *Geography Compass*, 7(5), 344–357.

Tseng, C. P. and Penning-Rowsell, E. C. 2012. Micro-political and related barriers to stakeholder engagement in flood risk management. *Geographical Journal*, 178(3), 253–269.

VanKoningsveld, M., Mulder, J. P. M., Stive, M. J. F., VanDerValk, L. and VanDerWeck, A. W. 2008. Living with sea-level rise and climate change: A case study of the Netherlands. *Journal of Coastal Research*, 24(2), 367–379.

Ward, P. J., Jongman, B., Weiland, F. S., Bouwman, A., van Beek, R., Bierkens, M. F. P., Ligtvoet, W. and Winsemius, H. 2013. Assessing flood risk at the global scale: Model setup, results, and sensitivity. *Environmental Research Letters*, 8(4), 44019.

Whatmore, S. J. 2009. Mapping knowledge controversies: science, democracy and the redistribution of expertise. *Progress in Human Geography*, 33(5), 587–598.

Wood, A. 2015. The politics of policy circulation: Unpacking the relationship between South African and South American cities in the adoption of bus rapid transit. *Antipode*, 47 (4), 1062–1079.

7

FLOOD POLICY PROCESS IN JAKARTA, INDONESIA, USING THE MULTIPLE STREAMS MODEL

Thanti Octavianti

Introduction

This chapter examines the usability of Kingdon's (2003) Multiple Streams (MS) model to understand the extent to which major flood events precipitated policy changes in Jakarta, Indonesia.

Scholars have often noted that major flood events have the ability to precipitate policy changes (Rosenthal and t'Hart, 1998; Johnson et al., 2005; Sultana et al., 2008). These disasters attract attention in policy communities (Birkland, 2006) because of the severe damages they cause. It may seem obvious that governments formulate policies as a response to a disaster (Tobin, 1997). However, such policies may be created in a rush and tend to 'subsidise some poor decisions and penalise some sound proposals' (Wilkins, 2000, 84). It would be different if the crisis acts as a catalyst to adopt a proposal that has been long discussed and carefully planned, when those in and around government work on the proposal prior to the crisis and utilise the crisis as a momentum to introduce change (Penning-Rowsell et al., 2006).

The MS model is widely used to analyse many policy processes. The model was developed in the context of health and aviation policies in the United States in the 1980s, but it has been used widely in different geographical and policy domains, such as in Argentine education policy (Jelena, 2008) and Canadian municipal emergency management policy (Henstra, 2010). Some authors (e.g. Meijerink, 2005) showed that the MS model can be combined with other policy theories, such as Sabatier's (1988) Advocacy Coalition Framework (see also chapter 8).

Focusing on the recent four major flood events in Jakarta (in 1996, 2002, 2007 and 2013), this chapter aims to analyse the extent to which the MS model can explain the policy processes and policy changes in relation to these floods. Located in a 5,000-year-old delta with more than 40 per cent of the area below sea level, Jakarta is naturally vulnerable to flooding (Caljouw et al., 2005). Urbanisation pressure exacerbates the flooding condition since it causes rapid floodplain development. The 2007 flood was arguably the worst flood in Jakarta's history. It was seen as a national calamity due to the extensive paralysis it caused. The total financial loss due to the flood was around US$ 900 million (Bappenas, 2007a) and threatened the global market outlook for Indonesia as a nation (Steinberg, 2007).

As a response, the Indonesian government enacted two important regulations relating to disaster management and spatial planning. This raises questions such as to what extent the regulations can be considered as true policy changes, and what factors, other than the flood, have contributed to these policy responses. Also for discussion is whether or not other major floods were able to trigger similar policy responses.

The Multiple Streams model

Derived from the garbage can model developed by Cohen, March and Olsen (1972), the Multiple Streams model developed by Kingdon (2003) is an analytical framework for agenda-setting and alternative specification. The model was first published in 1984 (Kingdon, 2003), and revised in 1995, 2003, 2011 and 2014. This chapter is based on the 2003 version, but insights from Kingdon (2014) have also been employed.

The main components of this policy process model are three independent streams: the problem stream, the politics stream and the policy stream. The problem stream refers to any conditions in the public domain that requires government action to address them. Conditions that are highlighted by indicators (numerical data) and focusing events are more likely to be brought to the attention of government officials than conditions that do not have those advantages of being quantified and supported by critical moments, such as floods.

The politics stream refers to a particular political condition in the jurisdiction, such as variations of national mood, administrative or legislative turnover, and interest group pressure campaigns. While political events have important policy consequences, not every political environment or event is equally significant in influencing a specific policy field.

The last stream, the policy stream, consists of proposed solutions developed by experts and analysts (Kingdon, 2003). Proposals are developed gradually in a so-called 'policy primeval soup', in which proposals are recast or combined with something else. Because of this slow progress of policy alternative development, the origins of them are – or can be – somewhat haphazard.

The three streams have lives of their own and interact when they are brought together at a critical juncture by the activities of policy entrepreneurs. This moment is what Kingdon terms 'coupling', when 'solutions become joined to problems, and both of them are joined to favourable political forces' (Kingdon, 2014, 194). The coupling of the streams opens a policy window that creates 'opportunity for action or given initiatives' (Kingdon, 2003, 166).

Kingdon notes that coupling is more than pushing as policy entrepreneurs must develop their proposals before the policy window opens. Windows are opened by events in either the problem or the political streams, creating problem windows and political windows, respectively. During a policy window, policy entrepreneurs attach their pet solutions to a problem that floats by (hence the 'stream' notion). The complete joining of all three streams dramatically enhances the odds that the subject will become firmly fixed on a decision agenda. The characteristics of these policy windows vary according to the streams' components (Howlett, 2009), but most of the time they are scarce and can stay open only for a short period of time.

The MS model highlights the importance of both policy process and policy entrepreneurs, but this chapter emphasises just the process element (the three streams and their joining to create a policy window).

Flooding in Jakarta through the Multiple Streams lens

In the past two decades, Jakarta has been annually exposed to flooding, with a severe flood happening almost every five years: 1996, 2002, 2007 and 2013. Table 7.1 summarises activities and events in the three streams – problem, politics and policy – in relation to these recent major floods.

The 1996 flood occurred in the dictatorial Suharto regime that strongly affected the policy and administration system at local levels. Funding for local governments was regulated by the central government and originated mainly from international donor agencies (Firman, 2008). It was aggravated by the unstable economic condition in the country due to the Asian financial crisis. Such political and economic unrest was unable to couple with the flooding problem to open a policy window. The fall of Suharto regime in 1998 started an era of democracy and the country adopted a new decentralised system. Jakarta also recovered from the financial crisis in early 2000. One of the impacts of this revived condition for flood policy was the enactment of local spatial planning regulations, known as Perda 6/1999. The 1996 flood was not able to open a policy window immediately after the flood, but it opened later when the situation was properly supportive.

In 2002, Jakarta was inundated by another severe flood. This problem interacted with a relatively stable political atmosphere in the new democratic era. Their coupling was able to open a policy window. A re-elected governor who acted as a policy entrepreneur was successful in advocating the most important flood alleviation project at that time: the East Flood Canal (EFC). The project was finally started in 2003 after 30 years of delay in implementation due to funding constraints and land acquisition problems (Simanjuntak et al., 2012).

Another policy window opened in 2007 after a further catastrophic flood (problem stream) combined with the first direct gubernatorial election (politics stream). As a result, new regulations of disaster management and spatial planning were enacted shortly after the flood. Furthermore, the newly elected governor utilised this window to initiate a dredging project in 2009. The government could not do any other high-cost flood defence projects in the city because the East Flood Canal was still under construction.

The most recent major flood, the 2013 event, happened just three months after another new governor was elected. A policy window opened because the flood in

TABLE 7.1 Key activities and events related to Jakarta's flood policy evolution since the mid-1990s

Flood event	Problem stream	Politics stream	Policy stream
Prior to 1996	• Major floods: 1621, 1654, 1918, 1976	1967 – Suharto's regime/ new order (dictatorship)	1973 – Flood Master Plan I 1991–1993 – Flood Master Plan II
January 1996	• The first major flood after 1976 • 745 houses affected • 2,640 evacuees	*Changing president:* • May 1998 –Suharto stepped down (end of dictatorship) *Election:* • Oct 1997 – New governor elected by parliament *Financial climate:* • 1997–1998 –Asian financial crisis • 2000 –Indonesia's positive economic growth *Regulation:* • UU★ 22/1999 Decentralised System Act (devolution of power) • UU 25/1999 Fiscal Equalization Act • UU 34/1999 Jakarta as Special District Act	• 1996 – New study about Jakarta's flood • 1997 – Flood Master Plan III
	Regulation: 1999 – Regional Regulation 6/1999 Jakarta's Spatial Planning		
February 2002	• 15–20% area inundated • 30 killed • 380,000 evacuees	*Regulation:* • UU 32/2004 Decentralisation Act (amendment) *Election:* • Sept 2002 –Incumbent governor re-elected by parliament	• 2002 – Flood Master Plan IV • 2005 – Jakarta's long-term plan (2005–2025), including flood risk management • 2002 – MoU to construct East Flood Canal (EFC)

		Implementation: *2003 – Instigation of East Flood Canal construction*
February 2007	• 60% area inundated • 70 people killed • 340,000 evacuees	*Election:* • Aug 2007 – First direct gubernatorial election • Jul 2012-Direct gubernatorial election	• 2007 – Proposal to build canal connecting River Ciliwung to East Flood Canal (EFC) • 2007 – Proposal to construct Multi-Purpose Deep Tunnel (MPDT) and Giant Sea Wall (GSW) to alleviate flood risk
	Regulation: *UU 24/2007 Disaster Management Law* *UU 26/2007 National Spatial Planning Law* *Implementation:* *2009 – River Ciliwung dredging project*		
January 2013	• 14% area inundated • 20 people killed • 50,000 evacuees	• New administration (flood happened three months after new governor elected in Oct 2012)	• Mar 2013 –MPDT and GSW included in the 2013 –2017 plan of Jakarta • Oct 2013 – MoU to construct Ciliwung Dam • Nov 2013 – The launch of the National Capital Integrated Coastal Development (NCICD) plan (seawall and reclamation projects) • Jan 2014 –Planning to construct a connecting canal between River Ciliwung and East Flood Canal (EFC)
	Implementation: *2013 – Normalisation of Pluit and Ria Rio retention ponds*		

Source: Bappenas, 2007a; Akmalah and Grigg, 2011; Simanjuntak et al., 2012; Vaswani, 2013.

Note: UU* stands for *Undang-Undang*, a national-level regulation in Indonesia.

the problem stream and the new administration in the politics stream came together. Since many proposals had 'floated' in the policy stream for more than two
decades, the new governor as the policy entrepreneur took advantage of the policy
window to implement some urgent projects (e.g. the normalisation of Pluit and
Ria Rio reservoirs). The window was also used to launch a massive seawall project
as part of the National Capital Integrated Coastal Development (NCICD) plan to
deal with the increased risk of flooding caused by land subsidence and sea level rise.

Implementation 'booster' and policy change

Implementation issues

Policy implementation deficit has characterised flood policies in Jakarta. There are
fractures between levels of government and organisation at the local level to implement the already adopted policy (Pressman and Wildavsky, 1984). The policy is
either implemented at a later date or sinks into oblivion and hence never gets
implemented. Common factors causing an implementation deficit are institutional
characteristics (Storbjork, 2007), lack of coordination between stakeholders (Beucher,
2009), unspecific mandate and insufficient resources (Montjoy and O'Toole, 1979).

As shown in the policy stream component in Table 7.1, the government actively
responded to recent flood events by undertaking studies and producing policies.
But there was a phenomenon of serious time lags between when a policy was
decided and it being implemented. Policy windows opened by major floods in
2002 and 2013 were able to implement these 'shelved' policies. We label this
particular outcome from a policy window an implementation 'booster'. The windows enable 'a renewed effort to implement an existing policy that had been
inadequately enforced' (Birkland, 2006, 189).

The implementation of the East Flood Canal – delayed for more than 30 years –
was finally instigated after the memorandum of understanding signed as a response to
the 2002 flood. That flood, supported by conducive political and financial conditions,
became a wake-up call for the government to act on the flooding problem. They
considered the construction of the canal was urgent and could significantly alleviate
the flooding in the city. This 23.6-kilometre canal was finally completed in 2010.

Another shelved policy and project that was finally implemented is the normalisation of Pluit and Ria Rio retention ponds. The project started after the 2013
flood, although these programmes had been on governmental agendas for about
ten years (Bappenas, 2007b). Normalisation aims to return the capacity of the
ponds to their design standards by dredging and clearing squatters from their banks.
The project was delayed because of the difficulties in negotiating with the people
who had inhabited the banks for decades. The government was reluctant to deal
with such a sensitive issue as it might have harmed a governor's popularity. The
2013 flood occurred just three months after a new governor took office. He
prioritised the normalisation project and used a persuasive, not repressive, approach
to discuss the issue with the people concerned.

Windows of opportunity and policy change

Kingdon (2003) does not define what is policy change; this is partly because he aims to explain policy process and does not specify what the outcome of policy windows should be. Therefore, we follow Johnson et al.'s (2005, 562) definition of policy change: 'changing set(s) of beliefs, values and attitudes towards the flood problem' because of learning from past disasters (Bennet and Howlett, 1992).

So, policy change means a shift of approach or thinking indicated by the enactment of law products containing a new idea. This appears to be a palpable definition, but since the essence here is the shift of ideas, the passing of a new law may or may not be called policy change depending on the ideas underpinning that change and its policy. Birkland (2006) further argues the need to adopt actual legislation as part of policy change definition. While we agree that it is necessary to implement the already enacted policy, we seek to separate the policy formulation and implementation stages.

The enactment of UU 24/2007 Disaster Management Act and UU 26/ 2007 Spatial Planning Act after the 2007 crisis can be considered as policy changes because they contain new approaches related to the existing flood risk management regime. The first act, UU 24/2007 Disaster Management Act, denotes a shift of approach from merely an emergency response to a disaster to an integrated approach of disaster management, emphasising the importance of early warning systems to reduce damage (UU 24/ 2007, article 44). The mandate consists of forming the National Disaster Management Agency and coordinating local disaster management strategies throughout the country (BNPB, 2013). The emergence of this new idea towards disaster management originated from the Tsunami catastrophe that struck Aceh in 2004 (Vidiarina, n.d.). Since then, the government had been actively developing a new regulation about disaster management strategy (BNPB, 2013) that was finally promulgated after the 2007 flood.

The general principles of the new spatial planning law in UU 26/2007 are still the same as the old ones, under UU 24/1992. However, the new act obliges local governments to provide a fixed amount of green spaces: 30 per cent of their total area (UU 26/2007, article 29). Such new standards force local governments to change fundamentally their planning strategies (Liesnawati, 2010). By having more open spaces, rainwater will infiltrate the ground instead of moving as surface runoff. The idea of requiring local governments to provide such spaces had been enacted in less binding regulations (e.g. Ministerial Regulation 327/KPTS/M/ 2002), but the momentum after the 2007 crisis could make this issue more impactful. Because Jakarta's crises are considered as national events, they can influence the regulations at the national level.

The policy formulation processes of each law indicate that the rate of policy changes was accelerated by a disaster. 'New' ideas carried in both laws did not emerge from the 2007 flood, but they had been developing from accumulative learning of past disasters. These seemingly 'new' ideas had been discussed and circulated in the widespread professional or public discourses (see Penning-Rowsell et al.,

2006). The 2007 flood acted as a catalyst to accelerate the enactment of regulations carrying pre-existing ideas that evolved through learning (see also chapter 9).

Lastly, neither policy change nor programme implementation occurred after the 1996 flood. The flood was not able to open a policy window immediately, but it opened three years later. The window did not open in 1996 because the problem stream was not coupled with the unstable political and economic conditions in the politics stream. In 1999, Regional Regulation 6/1999 about Jakarta's spatial planning was legalised. It was not clear whether this law contained new ideas for flood risk management, but it urged the implementation of the East Flood Canal construction and acted as a legal foundation for the instigation of the project in 2003. Nonetheless, due to the timing that policy enacted, it cannot be considered as a policy change that was precipitated by a crisis.

The usability of the Multiple Streams model

The MS model suggests that the problem and the politics streams interact first before policy stream joins to make the coupling. It is the 'change in the political stream … or a new problem capture[ing] the attention of governmental officials' that potentially opens a window (Kingdon, 2003, 168). In Jakarta's case, the interaction between these two streams was always initiated by a flood problem as a focusing event. Further, Kingdon (2003, 178) postulates that 'none of the streams are sufficient by themselves to place an item firmly on the agenda'. This was observed in the 1996 flood, in which a policy window did not open immediately after the crisis because it could not be coupled with the unstable political condition at that time.

This model also notes that a crisis 'merely opens a window of opportunity for change without guarantee of change itself' (Birkland, 2006, 24). This was evident in Jakarta as among the three windows opened, only the 2007 window was able to promote a policy change. This relates to the implementation deficit condition: the introduction of a new policy was not prioritised because relevant policies were still piled up to be implemented. Table 7.2 shows the relationship between the policy window, change and implementation in Jakarta's flood policy-making process.

Relating to the duration of a policy window, the MS model suggests that a window opens only for a short period of time. While this was observed in Jakarta, it appears that the window never closed completely because floods struck the city annually (see also chapter 8). Therefore, the flood problem was on the governmental agenda for most of the time. It was indicated by regulations produced by the government, although crisis did not occur (see the policy stream in Table 7.1). The severe floods, which occurred in a regular pattern, allowed policy windows to be opened longer than usual and were able to push the problem off the government agenda and onto the decision agenda and process.

Concerning the independent characteristic of the streams, Kingdon has also received criticism on this conceptualisation. Critics (e.g. Mucciaroni, 1992) argue that the streams tend to be interdependent because changes in one stream can

TABLE 7.2 Key characteristics of the Multiple Streams model

---

1 Problem stream

Conditions that are not highlighted by indicators, focusing events or feedback are less likely to be brought to the attention of government officials than conditions that do have those advantages.

---

2 Politics stream

Political events (e.g. change of administration) have important policy consequences, but not every environment or event is equally significant.

---

3 Policy stream

---

a. Policy alternatives are developed gradually in a 'policy primeval soup', akin to biological natural selection.
b. The origins of policy alternatives are somewhat haphazard.

---

4 Coupling

---

a. The separate streams of problems, policies and politics each have lives of their own.
b. Windows are opened by events in either the problem or political streams: problem windows and political windows.
c. Predictable or unpredictable, open windows are small and scarce and do not stay open long.
d. The complete joining of all three streams dramatically enhances the odds that subjects will become firmly fixed on a decision agenda.
e. Entrepreneurs have pet solutions and wait for problems to 'float by' to which they can attach their solutions.

---

trigger changes in another. In response to this, Kingdon, in his latest publication on the MS model (Kingdon, 2014), acknowledges the interdependency of streams to a certain extent: couplings are attempted often other than in open windows and final coupling. This interdependent notion was observed in the Jakarta's case, especially for the policy and problem streams. The ideas (policy stream) embodied in UU 24/2007 Disaster Management Act and UU 26/2007 Spatial Planning Act were products of learning from previous disasters (problem stream) and had been discussed in specialist communities long before the 2007 window opened. Before the enactment of these laws, the components of these two streams were clearly interacting.

While most of the elements of the MS model were evident in Jakarta's case, there is one observation that seems to be contradictory to the model's argument. The 'problem surfing' notion, in which solutions 'search(ing) for problems … to become attached' (Kingdon, 2003, 172) is not evident in this Jakarta case. Solutions or proposals that 'float' in the policy streams are emerging because of the problems. So, once a window of opportunity opens, they are attached to the resurfacing initial problem: flood crisis. The reverse notion in which a problem seeks whatever solutions are available is more appropriate in this context.

Given these insights from the Jakarta's case, we concur with other authors (John, 2000; Zahariadis, 2007; Robinson and Eller, 2010; Solik and Penning-Rowsell, 2017) regarding the importance of expanding the existing MS scope to include other stages of policy-making process: from only the agenda-setting and alternative specification stages to also include policy enactment and implementation stages.

Kingdon (2003, 4) sees the governmental agenda as 'a list of items that are getting attention', while decision agenda is 'a list of subjects within governmental agenda that are up for an active decision'. John (1999, 176) argues that Kingdon concentrates 'too much on agendas and not enough on how ideas feed into the implementation process and back again'. This limited scope of the MS policy analysis is the primary reason why our implementation 'booster' observed after policy windows is not recognised by the model, which excludes possibilities other than policy enactment as the expected result of the streams' coupling.

Conclusions

This chapter has demonstrated that the Multiple Streams model is a powerful lens through which to analyse flood policy process in Jakarta given its conceptualisation of three streams (the problem, politics and policy streams) and their convergence to create a policy window.

The case study shows that only when the three streams coupled had policy change a higher probability of occurrence, in which the political stream was the main determinant. From the four major floods (1996, 2002, 2007 and 2013), only the 2007 crisis acted as a catalyst to precipitate policy change in the city's flood policy by the legalisation of two regulations, ideas that had been developing for many years.

The 1996 flood was unable to open a policy window and created no change immediately after the crisis. The other major events, 2002 and 2013, acted as implementation 'boosters' to realise delayed policies. This 'booster' phenomenon is not recognised in the MS model because that model has the relatively limited scope of only explaining the initial stages of the policy process, primarily agenda setting and alternatives specification. Despite this limitation, the MS model offers a simple yet robust conceptualisation of policy formulation. The concepts of the three streams and policy window appear to be easily and usefully mobilised in a wide range of policy contexts to help explain the often chaotic and apparently irrational nature of the policy process.

References

Akmalah, E. and Grigg, N. (2011). Jakarta flooding: Systems study of socio-technical forces. *Water International*, 36(6), 733–747.

Bappenas (2007a) Damage and loss assessment report in Jabodetabek area caused by the early 2007 flood, Ministry of National Development Planning, Jakarta.

Bappenas (2007b) Flood alleviation strategy, Ministry of National Development Planning, Jakarta.

Bennett, C. and Howlett, M. (1992). The lessons of learning: Reconciling theories of policy learning and policy change. *Policy Sciences*, 25(3), 275–294.

Beucher, S. (2009) National/local policy tensions in flood risk management: An international comparison. *Environmental Hazards*, 8(2), 101–116.

Birkland, T. A. (2006). *Lessons of disaster: Policy change after catastrophic events*. Georgetown University Press, Washington, DC.

BNPB (2013). The National Disaster Management Agency. https://bnpb.go.id/, accessed 15 December 2013.

Caljouw, M., Nas, P. J. M. and Pratiwo, M. (2005). Flooding in Jakarta: Towards a blue city with improved water management. *Bijdragen tot de Taal-, Land- en Volkenkunde*, 161(4), 454–484.

Cohen, M. D., March, J. G. and Olsen, J. P. (1972). A garbage can model of organizational choice. *Administrative Science Quarterly*, 17(1), 1–25.

Firman, T. (2008). In search of a governance institution model for Jakarta Metropolitan Area (JMA) under Indonesia's new decentralisation policy: Old problems, new challenges. *Public Administration and Development*, 28(4), 280–290.

Henstra, D. (2010). Explaining local policy choices: A Multiple Streams analysis of municipal emergency management. *Canadian Public Administration*, 53(2), 241–258.

Howlett, M. (2009). *Studying public policy: Policy cycles & policy subsystems*, 3rd edition. Oxford University Press, Oxford.

Jelena, T. (2008). Why education policies fail: Multiple streams model of policymaking, *Zbornik Instituta za Pedagoška Istraživanja*, 40(1), 22–36.

John, P. (2000). *Analysing public policy*. Continuum, London.

Johnson, C. L., Tunstall, S. M. and Penning-Rowsell, E. C. (2005). Floods as catalysts for policy change: Historical lessons from England and Wales. *International Journal of Water Resources Development*, 21(4), 561–575.

Kingdon, J. W. (2003). *Agendas, alternatives, and public policies*, 2nd edition. Longman, New York and London.

Kingdon, J. W. (2014). *Agendas, alternatives, and public policies*, 2nd edition. Pearson Education, Harlow.

Liesnawati (2010). Implementation study of UU 26/2007 Spatial Planning Act in Semarang. Undergraduate thesis. University of Diponegoro, Indonesia.

Meijerink, S. (2005). Understanding policy stability and change: The interplay of advocacy coalitions and epistemic communities, windows of opportunity, and Dutch coastal flooding policy 1945–2003. *Journal of European Public Policy*, 12(6), 1060–1077.

Ministry of Housing and Infrastructure (2002). Ministerial Regulation 327/KPTS/M/2002: Six principles of spatial planning. Ministry of Housing and Infrastructure, Jakarta, Republic of Indonesia.

Montjoy, R. S. and O'Toole, L. J., Jr. (1979). Toward a theory of policy implementation: An organizational perspective. *Public Administration Review*, 39(5), 465–476.

Mucciaroni, G. (1992). The garbage can model and the study of policy making: A critique, *Polity*, 24(3), 459–482.

Penning-Rowsell, E. C., Johnson, C. and Tunstall, S. (2006). Signals from pre-crisis discourse: Lessons from UK flooding for global environmental policy change? *Global Environmental Change*, 16, 323–339.

Pressman, J. L. and Wildavsky, A. B. (1984). *Implementation: How great expectations in Washington are dashed in Oakland*, 3rd edition. University of California Press, Berkeley.

Putri, P. W. and Rahmanti, A. S. (2010). Jakarta waterscape: From structuring water to 21st-century hybrid nature? *Nakhara*, 6, 59–74.

Republic of Indonesia (2007a). *UU 24/2007 Disaster Management Act*.

Republic of Indonesia (2007b). *UU 26/2007 Spatial Planning Act*.

Robinson, S. E. and Eller, W. S. (2010). Participation in policy streams: Testing the separation of problems and solutions in subnational policy systems. *Policy Studies Journal*, 38(2), 199–216.

Rosenthal, U. and t'Hart, P. (1998). *Flood response and crisis management in Western Europe: A comparative analysis*. Springer, Heidelberg.

Sabatier, P. A. (1988). An advocacy coalition framework of policy change and the role of policy-oriented learning therein, *Policy Sciences*, 21, 129–168.

Simanjuntak, I., Frantzeskaki, N., Enserink, B. and Ravesteijn, W. (2012). Evaluating Jakarta's flood defence governance: The impact of political and institutional reforms. *Water Policy*, 14(4), 561–580.

Solik, B. and Penning-Rowsell, E. C. (2017). Adding an implementation phase to the framework for flood policy evolution: Insights from South Africa. *International Journal of Water Resources Development*, 33(1), 51–68.

Steinberg, F. (2007). Jakarta: Environmental problems and sustainability. *Habitat International*, 31(3), 354–365.

Storbjork, S. (2007). Governing climate adaptation in the local arena: Challenges of risk management and planning in Sweden. *Local Environment*, 12(5), 457–469.

Sultana, P., Johnson, C. and Thompson, P. (2008). The impact of major floods on flood risk policy evolution: Insights from Bangladesh. *International Journal of River Basin Management*, 6(4), 339–348.

Tobin, G. A. (1997). *Natural hazards: Explanation and integration*. Guilford Press, New York and London.

Vaswani, K. (2013). Indonesian capital Jakarta hit by deadly flooding. BBC News, 15 December, www.bbc.co.uk/news/world-asia-21054769.

Vidiarina, H. D. (n.d.). The legal framework of early warning in Indonesia. GTZ, Jakarta.

Wilkins, L. (2000). Searching for symbolic mitigation: media coverage of two floods. In Parker, D. J. (ed.), *Floods*. Routledge, London, 80–87.

Zahariadis, N. (2007). The Multiple Streams framework: Structures, limitations, prospects. In Sabatier, P. A. (ed.), *Theories of the policy process*, 2nd edition. Westview Press, Boulder, CO, 65–92.

8

A REVOLVING DOOR OF POLICY EVOLUTION

Climate change adaptation after Superstorm Sandy

Carey Goldman[1]

Introduction

There is a well-established connection between flooding disasters and subsequent demands from the public for policy change (Johnson et al., 2005). Much of the relevant literature focuses on this interplay of floods and flood policy, and it is clear that it takes an event of significant magnitude with severe impacts to influence the policy agenda (Parker, 2000) and a serious flood to trigger a range of policy responses (Johnson et al., 2005).

While much scrutiny has been given to this flood risk management (FRM) policy response after flooding, there is substantially less literature examining its impacts on policies beyond that field. As the impacts of climate change become more common and severe, a critical area of focus should be an analysis of those broader policy changes that might arise from serious flood events as an aid to developing more effective approaches to all such policy development in the future.

The Intergovernmental Panel on Climate Change (IPCC) argues that a high level of vulnerability across the socio-economic spectrum exists from a range of climatic extreme events – flooding being one of them (IPCC, 2012, 2014). These flood threats can include disrupting power supplies and the operation of critical institutions such as stock exchanges. Damage and destruction of private property, particularly in the most marginalised communities, can result from floods, but dislocation and loss can result from other sectors that are only indirectly flood-affected but interlinked (e.g. transportation). To take another example, threats to human health and life in severe floods often require emergency responses affecting health facilities and its infrastructure, influencing their capacity and its effectiveness in tackling the many non-flood health issues that they face. Given the significant impacts associated with extreme weather events, the need for a broad portfolio of interlinked climate adaptation programmes driven by appropriate climate adaptation policies is clear.

A graphic reminder of the increasing need to adapt came to the USA on 29 October 2012 in the form of 'Superstorm Sandy' (hereinafter simply 'Sandy'). Sandy was a very large storm that had devastating impacts along the east coast of the United States. Although direct attribution of a single event like Sandy to climate change is very difficult, long-term forecasting models used by the IPCC suggest that climate change will increase the frequency and intensity of such storms (IPCC, 2014). Additionally, the field of event attribution, while in its infancy, is now making the link between climate change and extreme weather events somewhat clearer (Trenberth et al., 2015). As a result, the need for climate resilience is growing (Ranson et al., 2014), and it is important to understand how extreme weather events, such as Sandy, affect the pace and nature of adaptation policy development generally.

Methods and definitions

To understand the impact of Sandy on the full range of public policies, it is important to be clear as to definitions and to contextualise our research within the commonly used academic frameworks of the policy process.

Thus, first, in this chapter 'adaptation' is defined as reducing vulnerabilities to climate change, whereas 'mitigation' applies to limiting the extent of climate change (typically through efforts to reduce greenhouse gas emissions). Second, the chapter uses the lens of political science ideas to analyse the policy evolution that has occurred in the aftermath of Sandy. Policy formation theories have been applied before to flood management in a climate change adaptation context (see Penning-Rowsell et al., 2017) but we aim here to apply these theories in an integrated or combined manner, building on its previous deployment (Johnson et al., 2005) and using information from three interviews of stakeholders in the post-Sandy situation (Table 8.1).

At the same time we present a novel interpretation that addresses the persistent and revolving nature of policy evolution to analyse how climate change adaptation policy across diverse sectors in the United States has evolved since Sandy. There is evidence – from the last five years – of significant change at all levels and this 'revolving door' framework provides a useful platform for understanding how an event like Sandy affects climate change adaptation policy and how it has also permeated other policy spheres.

Theoretical frameworks

At the foundation of many policy development frameworks is the examination of the causes of behavioural change, as this makes up the heart of these theoretical frameworks (Johnson et al. 2005). Johnson et al. (2005) argue that a single framework is insufficient to examine policy evolution due to the need to capture a range of mechanisms that drive human behaviour. As a result, this chapter analyses multiple frameworks and proposes a novel integrated or combined interpretation based

TABLE 8.1 Interview subjects and questions

A. *Kathy Baskin*
Director of Water Policy, Massachusetts Executive Office of Energy and Environmental Affairs
B. *Nathaly Agosto Filion*
Resiliency Manager, New Jersey Resilience Network
C. *Irene Nielson*
Climate Change Coordinator, Environment Protection Agency, Region 2 (New York and New Jersey)
Semi-structured interviews were conducted over both the telephone and using email. The questions used were:
1 What efforts were being taken on (climate) adaptation policy prior to Sandy?
2 Did Sandy have any impacts on either the policy process or attitudes towards adaptation?
3 What has been accomplished in terms of adaptation policy since Sandy?
4 Does the impact of Sandy still resonate in the political sphere?
5 Looking at current and future adaptation efforts, how do they differ from the trajectory of adaptation policy pre-Sandy?

on how different policy streams, human behaviour and catalytic events interact to create windows of opportunity for policy development. This analysis also indicates that these windows of opportunity exist for limited times and rely on the need for policy entrepreneurs and coalitions of actors to drive further evolution. Through this analysis it is possible to observe how Sandy impacted climate change adaptation policy both within the flood risk management (FRM) realm and more broadly.

One theoretical framework that is highly applicable to policy evolution post-Sandy is Baumgartner and Jones' Punctuated Equilibrium Theory (PET) (Baumgartner and Jones, 1993). PET is concerned with catalytic change – defined as an increase in the rate of policy change and by the extent of involvement by new actors who may then become 'policy entrepreneurs'. It is these newcomers that can change the status quo. Like PET, which looks at periods of catalytic change, Kingdon (1984) also examines 'windows of opportunity' (WOP) for periods of significant change.

Kingdon's policy streams model (see also chapter 7) is based upon policy evolution within the US federal system. In this model, Kingdon analyses how policy solutions evolve and how government and those with a stake in the policy perceive them. Kingdon argues that for these WOPs to be seized upon, there must be convergence between three main policy streams: the problem stream, policy stream and politics stream. The problem stream is born when attention is suddenly bestowed to a pressing issue, leading to short-lived windows of opportunity. The policy stream is derived from the 'primeval soup' of policy ideas that exist at any one time. The political stream is activated when policy-makers have sufficient motive paired with an opportunity to make change. This requires awareness of the problem coupled with an acceptable proposed solution. A fundamental point about the policy streams model is that all three streams act independently. When a WOP exists, change will only occur if these three streams are aligned (Kingdon, 1984, 2003, 2011).

The Advocacy Coalition Framework (ACF) is a simplification of Helco's (1974) ideas that macro-factors such as social, economic or political factors only account for a portion of policy change. The ACF's basic tenet is that to deal with the multitude of actors within a subsystem (interactive network of stakeholders, legislators, regulators, agencies, lobbyists), it is best to aggregate them into an advocacy coalition (Sabatier, 1988). Subsystems are defined by a distinct goal and geography with multiple stakeholders including government, interest groups, scientists etc, and in these settings allies are sought, becoming a coalition to promote through advocacy their shared interests (Sabatier and Weible, 2009). In this context, external perturbations are the main path for policy change and these can range from natural disaster to regime changes – anything that shifts agendas or the focus of the public and decision makers (Sabatier and Weible, 2009).

Integrating or combining aspects of these policy formation frameworks and applying them to the evolution of climate change policy post-Sandy provides analytical insights and is the focus of this chapter. Flooding events are often the catalyst for the punctuated equilibrium described in the PET model, creating windows of opportunity, which ultimately depend on advocacy coalitions to keep items related to policy evolution on the agenda after the WOP closes. Through a combined analysis of these frameworks, using the post-Sandy policy evolution as a test case, it is possible to reinterpret key components of these theories, identify areas where they fail to address or explain experienced policy evolution and develop a revised interpretive framework that better explains policy evolution after a severe event.

'Superstorm' Sandy

On 24 October 2012, Sandy was officially a Category One hurricane off the coast of Jamaica, but eventually peaked as a Category Three hurricane while still in the Caribbean. It went through several cycles of growth and made landfall in Jamaica and Cuba. However, by 29 October, as Sandy had moved north, cooler waters brought about a post-tropical cyclone stage categorisation that Sandy maintained when it made landfall in Brigantine, New Jersey, that same day (Blake et al., 2013).

While Sandy may have diminished in maximum wind speed after passing Cuba, the radial extent of the tropical storm grew and wind speeds doubled (Blake et al., 2013). Though Sandy was in a post-cyclone stage when making landfall in New Jersey, it was an expansive storm that was intensified by a coinciding coastal storm that brings strong winds from the northeast. The combination of these events spawned the popular reference 'Superstorm Sandy'.

Sandy damaged 650,000 homes and left 21.3 million people without power, which remained out for weeks to months in some areas of the US (Kunz et al., 2013). These losses exemplify the significance of the impacts, but they are just a fraction of the total damages. In the US alone, Sandy caused US$70.2 billion in damages (NOAA National Hurricane Center, 2017) and devastated areas in the Caribbean already greatly impacted by previous natural disasters, such as in Haiti.

Much of the damage inflicted by Sandy in the US resulted from storm surges that occurred across the entire east coast of the country, though they were most severe in the New York, New Jersey and Connecticut region (Braamskamp and Penning-Rowsell, 2018). Not only were there extensive instances of physical damage, lives were lost as well. When direct and indirect deaths for the United States are totalled, Sandy killed 159, with New York sustaining the highest number of deaths (Blake et al., 2013). As a result, Sandy became a significant catalyst for policy change.

Policy evolution: government actions, and attitudes and perceptions

Climate change adaptation policy emerged in the early 1990s and since then has transformed from an ambiguous theoretical concept to one that is rapidly gaining universal respect as a legitimate policy domain (Hansen et al., 2013). In the past, policies that dealt with adaptation have been implemented on an ad hoc basis as there was little top-down directive. However, this is changing: in the five-plus years since Sandy there have been specific policy changes as well as a change in attitude regarding adaptation polices, resulting in a more comprehensive and integrated approach to policy development. Due to the all-encompassing and interlinked nature of adaptation strategies, examining these policies covers a broad spectrum of sectors.

In the years following Sandy a range of policies and strategies emerged to address climate change adaptation. One year after Sandy, President Obama issued Executive Order 13653 'Preparing the United States for the Impacts of Climate Change'. This was the first national level directive on climate change adaptation. This action made it clear that Obama and the EPA put climate change at the top of the policy agenda (Kellog, 2013). In the US EPA's 2013 Climate Change Adaptation Plan, Sandy was referenced nearly 50 times, highlighting the need to implement adaptation strategies in utilities/services such as water, power, and waste treatment. The US Federal Emergency Management Agency (FEMA) issued a policy directive to increase the Agency's adaptation efforts through research, improved building regulations, and flood insurance. However, in late 2017 President Trump eliminated the FEMA Federal Flood Risk Mitigation Standards. In addition to the withdrawal from the Paris climate accord this demonstrated the recent efforts to stall or backtrack on previous progress in climate policy.

At the city level, however, the rate of policy development was considerably more rapid, although states' and regions' commitment to and pace of evolution of adaptation strategies varied. As might be expected, the pace and extent of policy development was greatest in those locations most affected by Sandy. For example, New York City announced a US$19.5 billion plan to address resiliency within the city within a year of the storm. More recently, in 2017 the city announced new Climate Resiliency Design Guidelines to incorporate compressive adaptation strategies for all public works, in addition to new resilience projects, demonstrating that the appetite for climate adaptation policy evolution has been sustained in this region, a stark contrast with the federal policy agenda (Cunniff, 2017, 2018).

One of the most influential impacts of Sandy is the change in attitudes regarding climate change adaptation. Interviewee Irene Nielson the Climate Change Coordinator for New York and New Jersey of the USA's Environmental Protection Agency's (EPA) stated, 'Nationally, Superstorm Sandy helped propel adaptation policy…'. This notion of adaptation being thrust to the top of policy agendas both nationally and regionally was reiterated often. Sarah Kellog writes of the influence Sandy had in the Washington Lawyer (Kellog, 2013, 1): 'When Hurricane Sandy tore through the East Coast in October 2012, it did for public perceptions of climate change what no number of scientists, years of advocacy, millions of dollars in lobbying, or decades of peer–reviewed research could do, it made extreme weather and climate change a real and present threat'.

The key lesson from Kellogg's statement is the idea of a real and present danger. This idea that Sandy was a resounding wake-up call to the actualities of climate change appeared repeatedly – in opinion pieces, editorials and general conversation amongst the public. Sandy changed attitudes from treating climate change as an abstract theory and adaptation as a distant requirement to seeing climate change as a present reality: accepting that our world is changing and we must adapt now. Echoing this sentiment, interviewee Nathaly Agosto Filion, a resiliency manager with the New Jersey Resiliency Network, stated, 'Whereas climate adaptation was previously an exercise in looking at the future, it is now considered part of due diligence … a part of our current world'. This again exemplifies the transformative effect Sandy had on not only public perception of climate change but also its realm in the political sphere.

In addition, we see evidence of the more comprehensive and integrated approach to policy in the aftermath of Sandy. Filion explained that all planning in New Jersey, from housing to transportation, must now have a resiliency component, something inconceivable prior to Sandy. Although at the time of these interviews it had been two years since Sandy, there was still tangible evidence of how it changed people's view of climate change adaptation. As Filion concluded, 'Sandy taught me a lot about vulnerabilities'. Much of the cascading impacts, such as impacts on energy infrastructure were unanticipated (Kellog, 2013). Though climate change's role in Sandy is unknowable, the resulting damages made it impossible to deny the USA's vulnerabilities to large storms, and made more believable scientists' claims of a relationship between climate change and storm severities.

Policy evolution cannot exist without public support and pressure (Harries and Penning-Rowsell, 2011), and changing the public's perception regarding the need for climate adaptation drives policy evolution. Vicki Arroyo, executive director of the Georgetown Climate Canter at Georgetown University Law, explained (see Kellog, 2013), 'We're seeing an interest at the state level in adaptation and resilience, even in states that are not progressive on climate policy'. Sandy's influence on the public's perception of adaptation catapulted it to the top of political agendas even in areas that were previously resistant to mitigation efforts. As Arroyo and Nielson indicate, post-Sandy adaptation policy began to evolve in multiple ways and novel arenas.

This new and reframed focus on adaptation is clear, as Filion observed, 'Making the case for adaptation [is] easier post Sandy'. In some cases, existing organisations have reframed their objectives around adaptation to align them with current priorities. Now adaptation planning has infiltrated spheres across the socio-economic spectrum. Accordingly, we see resilience planning requirements, funding for adaptation projects, and new types of adaptation planning organisations being integrated into policy development at both federal and regional level.

In this vein, the EPA aimed to increase its resiliency planning, both internally and externally. This was streamlined by incorporating resilience requirements into existing planning processes. An example of this is found in brownfield grants aimed at the redevelopment of contaminated industrial waste sites as new housing, parks and other renewed use. Starting in 2013, the EPA required those applying to its Cleanup and Revolving Loan Fund grants to 'evaluate the resilience of the remedial options in light of reasonably foreseeable changing climate conditions (e.g. sea level rise, increased frequency and intensity of flooding and/or extreme weather events, etc.)' (EPA, 2014, 1), indicative of how Sandy has allowed adaptation policy evolution to permeate beyond FRM.

New types of funding mechanisms, or the integration of resiliency into existing funding mechanisms, is another area of policy evolution post-Sandy. An example of this is the Energy Resilience Bank, an innovative funding mechanism that provides finance for enhancing resiliency in the energy sector in the region Sandy hit hardest. Another example is FEMA's use of a Community Rating System (CRS) in its flood insurance to encourage increasing resiliency above required levels.

Another outcome of EO 13653 was the formation of the President's State, Local and Tribal Leaders Task Force on Climate Preparedness and Resilience, which reported in 2014 to provide top-down guidance regarding climate change adaptation for regional, state and local authorities that had been previously absent. However, as mentioned above, the current federal administration is actively working to reverse some of the climate change adaptation policies adopted during the Obama administration, an indication of how political cycles can affect policy evolution differently at different levels.

One component of New York City's response to Sandy was the development of the Sandy Regional Assembly (SRA), composed of 40 community-based non-government organisations (NGOs) that developed a 'recovery agenda' that explicitly outlined community-led adaptation as a main goal (Cunniff, 2017, 2018). Agosto Filion indicated in our interview that the increase in actors becoming involved with adaptation was complex; while there were undoubtedly new players, there were also many existing actors who have subtly reframed their arguments around adaptation and resiliency.

Another example of action at the city level is in Houston, where even a city with major fossil fuel interests has demonstrated a commitment to adhering to the Paris Climate Accord and has done much to improve the amount of renewable energy the city uses, although addressing climate change adaptation directly has been less of a priority, despite leading the US in casualties and monetary damage due to flooding (Dart, 2017). However, there are local groups aiming to elevate adaptation on the policy agenda. For example, an advocacy group called Residents Against Flooding have filed a lawsuit against the city to improve design standards and reduce flooding risk (Dart, 2017). Other stakeholders, such as a researcher at Texas A&M University, have developed tools for residents independently to assess the flood risk of their properties to supplement lack of formally available resilience tools and policies (Dart, 2017).

The Obama era supported a groundswell of policy development at the regional level – and unlike at the federal level, this regional policy development continues. Over 240 mayors across the US have pledged to uphold the Paris Climate Accord, as have many major corporations. At the regional level one can see sustained climate adaptation policy activity in locations in the United States that were significantly impacted by the 2017 hurricane season. Of note, areas of Texas, Florida, and Puerto Rico were particularly devastated by hurricanes Harvey, Irma and Maria respectively. Just as Sandy provided a catalyst for change in the Northeast region, as these regions face the realities of dealing with major flooding worsened by climate change, they continue their efforts to adapt and become more resilient. For example, the South Eastern Florida Regional Climate Change Compact is a partnership involving multiple counties, government agencies and researchers, which released their second Regional Climate Action Plan (RCAP) in December 2017 with an emphasis on resilience, particularly around marginalised sectors. This provides leaders in local government, farmers, NGOs and business leaders with more tools to implement improved adaptation practices and standards (Center for American Progress, 2018).

Coming back specifically to the Northeast region, overall, the direct impacts of Sandy on the region have driven a rapid and significant increase of involvement by government agencies, stakeholders, NGOs, organisations and the business sector in climate change adaptation. All of these actors have played a significant role in the evolution of climate change policy since Sandy. That evolution has included new planning, funding and development policies and programmes at federal, state, local and regional level. That evolution has sustained but slowed somewhat over the five-plus years since Sandy and, in the case of the federal government, has started to reverse. This same pattern of sustained climate adaptation policy activity at the city and regional level can be seen in areas impacted by other major storms post-Sandy.

Discussion

It is clear that Sandy had a significant influence in the evolution of US climate change adaptation policy in the five years since the event. After the storm there was an acceleration in adaptation policies at all levels of government. As the punctuated equilibrium theory (PET) suggests, this is the result of a change in status quo driven by policy entrepreneurs (Baumgartner and Jones, 1993). Sandy undoubtedly brought new actors into the adaptation sphere, a prerequisite and driver of catalytic change in the PET model. Coalitions formed in response to Sandy, as the ACF suggests will occur when complex issues involve many actors. There are multiple efforts to unite various factions of industries and NGOs impacted by Sandy and typically the most common ground is found vis-à-vis adaptation.

Of course, adaptation policy was not pulled from thin air in the wake of Sandy. Adaptation policy had been brewing in the primeval soup of policy options, as Kingdon's theory suggests (Kingdon, 1984; Penning-Rowsell et al., 2006). In dramatically changing the public's perception on the necessity of climate change adaptation, Sandy forced the first step of pushing adaptation policy evolution by igniting the public stream. Surveying done across coastal New England by MIT found that extreme weather events were the leading cause in shifting people's outlook on the possibility that climate change will impact their towns (NECAP, 2014). Subsequently, the policy and problem streams aligned with the politics stream, as Sandy exposed significant vulnerabilities and thus a window of opportunity was opened for substantial policy changes.

There are stark differences in policy outcomes between regional and federal regions. While elements of the PET and ACF were met at the federal level through various immediate agency-wide task forces and plans, the evolution of regional-level policies has been more aligned with these models over the full five years since Sandy. On the federal level, the change in administration was a clear disruption to the policy stream. As opposed to acting as a catalyst for policy evolution, the current administration caused a stoppage and has attempted backtracking where possible when it came to climate change adaptation policy on a federal level.

While the frameworks used in this study provide a critical lens to analyse policy evolution, they are not without their pitfalls. The windows of opportunity (WOP) theory states that there is a finite amount of time when all three streams (the problem stream, policy stream and politics stream) can be in unison and policy evolution can take place in the aftermath of a catalytic event. We would argue, however, that this short timeframe on windows of opportunity is overstated and that these windows of opportunity open and close on a periodic basis. As the most recent federal administration has shown, it is even possible to reverse these outcomes. It also shows that policy evolution happens at different paces (and perhaps different directions) at different levels of government.

Five years after Sandy there are numerous examples of the storm still impacting policy. While two of our interviewees hinted that Sandy's influence was slipping slightly by the end of year four, it is clear that work continues at both national and regional level. Regionally, Sandy lingers as people are still recovering from the damage and many policy changes directly tied to Sandy are still being implemented. Thus, the notion of a short-term WOP for policy change is not consistent with actual events post-Sandy. A more appropriate metaphor for Kingdon's theory would be a revolving door.

With the WOP theory, an event opens a window for change, which is then quickly shut. Instead of thinking in static terms of open or closed, a revolving door expresses the continuality and variability of these processes. When an event happens, one part of the revolving door opens, before shrinking and becoming closed off again. Subsequently, perhaps with an additional catalyst, the revolving door opens again, partially or fully. Periods of catalytic change occur when the door is wide open, incremental changes occur when the door is closed or part-closed. This revolving door metaphor more accurately captures, we believe, the entire cumulative arc of climate change adaptation policy evolution, even the most recent rollbacks at the federal level (after all, a revolving door spins in two directions) by acknowledging it is a transitional process as opposed to a static 'black or white' opportunity for progress.

In the context of climate change adaptation policy this metaphor is more appropriate because it takes into account the cumulative impact of previous events, current trends and impacts. Additionally, thinking of policy evolution as a revolving door rather than a window is more nuanced than simply open or closed, and is more consistent with the varying timeframes and levels of policy responses in the aftermath of Sandy and other storms. Additionally, this metaphor is well suited for catalytic events as they often slowly fade from thought till the next major event. With each event, and turn of the door, policy evolves, building on each rotation.

Conclusions

Policies evolve in a variety of ways: this is reflected in the range of theoretical frameworks that exist to analyse these changes. Understanding frameworks for policy change and applying them in an integrated or combined manner to the evolution of policy post-Sandy provides useful insights. The application of concepts from multiple frameworks, along with a critical analysis of how those frameworks align with reality, results in meaningful conclusions about the nature of climate change adaptation policy evolution in the aftermath of Sandy.

First, Sandy was a dramatic and certain catalyst for policy change, resulting in significant increases in policies promoting climate change adaptation. This chapter shows that these policy advances permeated not just the FRM realm but were evident in a broad range of policy domains. Second, the regional response to Sandy provided evidence of alignment with key elements of policy change frameworks, including the accelerated change associated with PET, and the formation of the coalitions anticipated by the ACF. Lastly, there were elements that aligned with the limited timeframe for change suggested by the WOP theory, but closer analysis concluded that the binary nature of these models (i.e. a window open or closed) may be overstated.

Accordingly, we have proposed the revolving door metaphor to capture the evolving nature of climate change adaptation policy. Applying historical frameworks for policy evolution to adaptation in a post-Sandy world thus has yielded not only interesting findings on Sandy's impact on adaptation policy, but also insightful critiques of these frameworks themselves.

Note

1 ORCID: 0000-0002-9092-747X

References

Baumgartner, F. R. and Jones, B. D. (1993). *Agendas and instability in American politics*. Chicago: University of Chicago Press.

Blake, E., Kimberlain, T., Berg, R., Cagialosi, J. and Beven, J. (2013). Tropical Cyclone Report: Hurricane Sandy, 22–29 October 2012. NOAA National Hurricane Center, Miami, FA.

Braamskamp, A. and Penning-Rowsell, E. C. (2018). Managed retreat: A rare and paradoxical success, but yielding a dismal prognosis, *Environmental Management and Sustainable Development*, 7(2), 108–136.

Center for American Progress (CAP) (2018). Social equity key to southeast Florida RCAP 2.0. CAP, Washington, DC.

Cohen, M. D., March, J. G., Olsen, J. P. and Morales, G. R. (2011). A garbage can model of organizational choice. *Gestion y Politica Publica*, 20(2), 247–290.

Cunniff, S. E. (2017). Exploring FEMA's Community Rating System as a tool for improving flood hazard mitigation and use of natural infrastructure: Initial summary of an EDF emerging issues workshop, August 9–10, 2017. Environmental Defense Fund, Washington, DC.

Cunniff, S. E. (2018). Improving FEMA's Community Rating System to encourage investment in coastal natural infrastructure to reduce storm damages, *Shore & Beach*, 86 (2), 27–32.

Dart, T. (2017). Houston fears climate change will cause catastrophic flooding. *The Guardian*, 16 June.

Denzin, N. K. and Lincoln, Y. S. (eds) (2013). *Collecting and interpreting qualitative materials*. 4th edition, Thousand Oaks, CA: Sage.

Department of Housing and Urban Development (2013). Hurricane Sandy rebuilding strategy. US Department of Housing and Urban Development, Washington, DC. Available at: www.hud.gov/sites/documents/HSREBUILDINGSTRATEGY.PDF

Environmental Protection Agency (2014). How to address changing climate concerns in an analysis of brownfield cleanup alternatives (ABCA). EPA, Washington, DC.

Hansen, J., Kharecha, P., Sato, M., Masson-Delmotte, V. et al. (2013). Assessing 'dangerous climate change': Required reduction of carbon emissions to protect young people, future generations and nature. *PLOS ONE*, 8, e81648, doi:10.1371/journal.pone.0081648.

Harries, T. and Penning-Rowsell, E. C. (2011). Victim pressure, institutional inertia and climate change adaptation: The case of flood risk. *Global Environmental Change*, 21(1), 188–197.

Helco, H. (1974). *Modern social policies in Britain and Sweden: From relief to income maintenance.* New Haven, CT: Yale University Press.

IPCC (Intergovernmental Panel on Climate Change) (2012). *Managing the risks of extreme events and disasters to advance climate change adaptation: Special report of the Intergovernmental Panel on Climate Change.* Cambridge: Cambridge University Press.

IPCC (Intergovernmental Panel on Climate Change) (2014). Summary for policymakers. In: *Climate Change 2014: Impacts, adaptation, and vulnerability. Part A: Global and sectoral aspects. Contribution of Working Group II to the Fifth Assessment Report of the Intergovernmental Panel on Climate Change.* Cambridge: Cambridge University Press.

Johnson, C. L., Tunstall, S. M. and Penning-Rowsell, E. C. (2005). Floods as Catalysts for Policy Change: Historical Lessons from England and Wales. *International Journal of Water Resources Development,* 21(4), 561–575.

Kellog, S. (2013). The cost of doing nothing. *Washington Lawyer* [online]. Accessed at: www.dcbar.org/bar-resources/publications/washington-lawyer/articles/may-2013-hurrican-cost.cfm.

Kingdon, J. W. (1984). *Agendas, alternatives, and public policies.* Boston: Little, Brown.

Kingdon, J. W. (2003). *Agendas, alternatives, and public policies.* 2nd edition. London: Longman.

Kingdon, J. W. (2011). *Agendas, alternatives, and public policies.* Updated 2nd edition. London: Longman.

Kunzi, M., Mühr, B., Kunz-Plapp, T., Daniell, J. E. et al. (2013). Investigation of super-storm Sandy 2012 in a multi-disciplinary approach. *Natural Hazards and Earth Systems Science,* 13, 2579–2598.

NOAA (National Oceanic and Atmospheric Administration) National Hurricane Center (2018). www.nhc.noaa.gov/data/tcr/.

NECAP (New England Climate Adaptation Project) (2014) New England Climate Adaptation Project, Massachusetts Institute of Technology & Consensus Building Institute (MITCBI), Cambridge, MA. Accessed at: https://necap.mit.edu/public-polls.

Parker, D. J. (ed.) (2000). *Floods.* Routledge, London.

Penning-Rowsell, E. C., Johnson, C. and Tunstall, S. M. (2006). 'Signals' from pre-crisis discourse: Lessons from UK flooding for global environmental policy change? *Global Environmental Change,* 16, 323–339.

Penning-Rowsell, E. C., Johnson, C. and Tunstall, S. M. (2017). Understanding policy change in flood risk management. *Water Security,* 2, 11–18.

Ranson, M., Kousky, C., Ruth, M., Jantarasami, L., Crimmins, A. and Tarquinio, L. (2014). Tropical and extratropical cyclone damages under climate change. *Climatic Change,* 127 (2), 227–241.

Sabatier, P. A. (1988). An advocacy coalition framework of policy change, and the role of policy-oriented learning therein. *Policy Sciences,* 21 (2–3), 129–168.

Sabatier, P. A. (2009). *Theories of the policy process.* 2nd edition. Boulder, CO: Westview Press.

Sabatier, P. A. and Weible, C. M. (2009). The advocacy coalition framework: innovations and clarifications, in Sabatier, P. A. (ed.), *Theories of the policy process.* Boulder, CO: Westview Press.

Sandy Regional Authority (2013). Recovery agenda. Recovery from the ground up: Strategies from community-based resiliency in New York and New Jersey. Sandy Regional Authority, New York City.

Trenberth, K. E., Fasullo, J. T. and Shepard, T. G. (2015). Attribution of climate extreme event. *Nature: Climate Change,* 5, 725–730.

9

POLICY BELIEF CHANGE AND LEARNING IN RESPONSE TO CALIFORNIA FLOODING

Clarke A. Knight

Introduction

Among California's vast resources, water has been both a creator and a destroyer. On the one hand water has transformed the state into one of the world's agricultural epicentres, delivering wealth and opportunity. On the other hand, repeated flooding has caused devastation, delivering painful lessons about river control and water management. Water's duality has raised politically sensitive management questions regarding best practices for the millions of urban users, farmers and wildlife that depend on its flows.

This chapter investigates the genesis of modern Californian flood policy, starting with the acute 1986 flood event, to understand how competing interests have intertwined to shape water policy in California. In the twentieth century, massive flood events motivated a centralised, engineering approach to water management that prevailed until the early 1990s, following those devastating 1986 floods. The question we raise is how did the major shift towards ecologically focused flood management occur in the two decades after 1986, given that previous major floods did not provoke such change.

The advocacy coalition framework (ACF) of policy change is well suited to modelling long-term policy change in California's flood management. The ACF profitably addresses contentious public policy problems when the divisions are deep and long-term (a decade or longer) (Weible and Sabatier, 2006). The ACF relies on the concept of coalitions containing many actors and the concept of policy subsystems at multiple levels. It stands in contrast to more traditional theoretical models that attribute policy change to 'dramatic events or crises, changes in governing coalitions, and administrative and legislative turnover' (Schlager, 2007, 310).

We examine here the major flooding events and early flood policy between 1986 and 2005. This examination first establishes the state's policy status quo before the California Department of Water Resources (DWR) was created. Next, the DWR's policy core beliefs in its early years are explained, followed by an outline of the simultaneous development of important policy coalitions. The immediate impact of the 1986 floods gives way to an exploration of evidence showing policy core belief change in the DWR from human-centric to an integrated human-ecosystem approach. The chapter concludes with an analysis of why substantial policy change (policy-oriented learning) occurred after 1999.

The advocacy coalition framework

The ACF was developed to better understand intractable public policy problems (Sabatier and Jenkins-Smith, 1999), and most of its applications concern environmental and resource policymaking, including flood policy case studies (e.g. Meijerink, 2005; Albright, 2011). The ACF unites actors with common policy beliefs into 'advocacy coalitions' (Sabatier, 1987), which work together in a continuous feedback loop of information, external and internal forces, and the dynamics acting on them (Weible and Sabatier, 2006). Coalitions from diverse institutions form a subsystem whereby policy change might emerge (Schlager, 2007). Policy subsystems are broad. They include 'journalists, researchers, and policy analysts who play important roles in the generation, dissemination, and evaluation of policy ideas' (Jenkins-Smith and Sabatier, 1993, 179), in addition to the traditional 'iron triangles', i.e. the legislative actors, administrative agencies and interest groups on one governmental level.

Within the ACF, policy-oriented learning is one mechanism causing policy change. Policy-oriented learning (POL) is the 'enduring alterations of thought or behavioural intentions' (Sabatier 1993, 42) that alters the beliefs of actors in the policy subsystem over a decade or longer. Individuals in subsystems learn, allowing new beliefs and attitudes to spread among many individuals (Weible and Sabatier, 2006). Albright's 2011 study of Hungarian flood policy illustrates this process (Albright, 2011).

In addition to POL, beliefs are central to the ACF and are broken into three levels. First are the 'deep core beliefs' that maintain the coalition's 'basic normative commitments and causal perceptions across an entire policy domain or subsystem' (Jenkins-Smith and Sabatier, 1994, 180); these are hypothesised to be enduring. The second degree of beliefs are the 'policy core beliefs' that are shared within a coalition. These values can change with major perturbations, such as a national flood (Sabatier, 1999), and are hypothesised to have changed in this case study. 'Secondary beliefs' – the least tightly held positions – may change without major perturbations because they relate to specific policy proposals that already align with core beliefs (Weible and Sabatier, 2006). For instance, a coalition of environmentalists (environmentalism is a deep core belief) may hold anti-development positions (policy core beliefs) and advocate for specific policies concerning endangered species (e.g. increasing wilderness areas, a secondary belief).

The ACF can be a powerful lens through which to examine the policy process under certain circumstances. One of its key strengths lies in its acknowledgment of individual psychological factors, that is that individuals have deeply held beliefs that shape their fundamental policy positions and those of the institutions in which they work (Weible and Sabatier, 2006). Other policy frameworks neglect the psychological biases of scientists, analysts and other actors. Additionally, the concept of POL provides a pathway to policy change when certain theoretical and operational conditions are met, though it has not been applied to many case studies.

Using components of the ACF designed to explain long-term change – POL and policy core beliefs – the development of California's flood management policy can be examined systematically. California's commitment to integrated water management (Biswas, 2004) and ecosystem-centric values is evident in current local and state policy (DWR, 2018), but this was not always the case. POL is detectable when coordinated activities among the various actors and institutions resulted in instances of changes in flood-regulated policies or programmes, following methods from Albright (2011) and Sabatier and Jenkins-Smith (1993, 1999). Evidence for the presence of policy core beliefs is found in state and federal laws and policies, personal communications, flood-related documents (especially the state water plan updates), professional forums, media publications and secondary literature.

Early Californian flood policy (1850–1956)

Since 1850, engineering solutions have been used to control the state's water, particularly in the fertile Central Valley, via a complex system of dams, reservoirs, levees, bypasses, weirs and other structures on nearly every watercourse. State-federal flood protection projects began in earnest under the Flood Control Act of 1917 after the California Debris Commission report in 1913 illustrated the need for weir construction and the channelising of rivers. Significant flooding events highlighted the need for the 1917 Act. For instance, the St. Francis Dam collapsed in 1928, killing 450 people in a flood wave. Heavy rains across southern California in 1938 caused floods that killed dozens and caused massive damage. This destruction led Congress to extend the Act in 1937 and 1941, authorising the US Army Corps of Engineers (hereafter the Corps) to start channelising the Los Angeles River (Kelley, 1989).

These laws represent the construction-oriented culture not only in government but also in the grassroots at that time (see also chapter 7). For instance, levees were constructed by local people before governmental authorisation by the 1917 law that allowed nearly 1,000 miles of construction (1,600 km). Local support, too, was brought in through the 1937 and 1941 acts which stipulated that the county provide rights-of-way as well as hold the federal government free from flood damage claims. By requiring local cooperation in flood defence, the government began to shift responsibility to local agencies and therefore seeded the grassroots coalitions of modern water policy. These extensions provided more funding and expanded development to include the San Joaquin Valley, illustrating the demand for these protections (DWR, 2010).

The policy core beliefs of the DWR (1950–1983)

The catastrophic 1955 Yuba City Flood, which destroyed Yuba city and forced the evacuation of over 30,000 people (Yuba City Water Agency, 2012), prompted the DWR's creation by Governor Knight. The engineering mindset from the early twentieth century thus became enshrined into the policy core beliefs of formal flood management activities in California from the beginnings of the DWR in 1956; this did not change for decades. In response to Yuba, the DWR's first director, Harvey Oren Banks, released the inaugural 1957 bulletin that highlighted the human-centric core beliefs of the agency. Banks stated that the DWR's mission was the 'control, protection, conservation, distribution, and utilisation of all the waters of California, to meet present and future needs for all beneficial uses and purposes in all areas of the state to the maximum feasible extent …. The full solution of California's water problems thus becomes essentially a financial and engineering problem' (DWR, 1957, XIII). 'Beneficial uses' meant agricultural and industrial water use, following the 1937 and 1941 policies (described above), as opposed to 'wasteful' water usage on wildlife. The goal of policy was to protect agricultural land for food and economic security, as well as rural flood defence. Banks' powerful statement demonstrates the policy core belief of human domination over nature, with development as the vehicle.

As the lead agency for flood-related activities, the DWR assumed responsibility for flood preparation, flood forecasting, real-time flood management and coordination with local agencies (Taylor, 2017). These required feats of engineering to control the massive rivers across California for flood defence, agriculture and growing urban centres. Therefore a construction mentality endured over several decades. Because the DWR writes the official state water plan, the official policy positions of the DWR and their policy core beliefs were and remain explicit. The DWR's stated mission therefore illustrates their deep core belief that humans can (and must) control nature by continuing development, a policy core belief (summarised in Table 9.1) A slight change in the stated mission occurred in 1974 with the inclusion of language about 'subjective factors' (DWR, 1974, 15), but these aspects were not given serious consideration within the report, which suggests that policy core beliefs were stable.

Coalition development

In addition to the powerful agricultural coalition, which had maintained its pro-construction stance and been aided by twentieth-century food security policies, some urban water users also expressed pro-development tendencies after flooding events. In 1960, voters authorised $1.75 billion in bonds to construct a new water conveyance system, with the secondary aim of state-wide flood risk management. Voters approved the California State Water Project in 1967, which allowed for the construction of the Oroville Dam to control the last major tributary of the Sacramento River, the Feather River (Kelley, 1989). With this dam, the Corps

TABLE 9.1 Stated DWR goals with corresponding deep core beliefs and policy core beliefs

Update year	Stated DWR mission	Deep core beliefs	Policy core beliefs
1957	Growth, solving engineering problems	Human domination of nature	Pro-development
1966	Development, building structures	Human domination of nature	Pro-development
1970	Development, some environmental concerns	Human domination of nature	Pro-development
1974	Finding a balance between economic and 'subjective factors'	Slight backing away from human domination of nature	Pro-development
1983	Self-defined as a technical report and user's manual for water development	Human domination of nature	Pro-development

completed their long-term goal of large headwater dams in the Sacramento and San Joaquin Valleys, satisfying the agricultural coalition which had been lobbying for increased water allocation.

The rise of a new preservationist coalition, based on the idea that water must be shared with wildlife, is first detectable in the 1970s. This coalition holds a fundamentally different position to that held by the agricultural coalition or urban water users' coalition, opposing the notion that humans dominate nature. Their popularity was evident in communities. For instance, when the Corps twice proposed project designs in the 1970s to control flooding in the Napa Valley, both designs were rejected. Voters cited financial and, importantly, environmental reasons (Ferguson, 2017). In fact, the 1960 and 1970 environmental debates happening across the country have been tied to the shift from technocratic planning to planning that included public participation and wider stakeholder involvement (Loh, 1994). Thus, three main coalitions, comprising urban water users, farmers and environmentalists, jockeyed over the state's water in hearings, courtrooms and at the ballot box for much of the 1990s and 2000s (DWR, 1993; Ambruster, 2008). As coalition representatives took part in DWR planning committees (discussed below), their competing interests influenced the direction of California's water policy to an ever-greater degree.

The historic February 1986 floods had several immediate impacts. First, the storm was physically devastating. Over 50,000 people were forced to leave their homes, and the damages totalled some $500 million. The DWR called it the greatest storm on record (DWR, 1986). Second, the floods brought the 'Big Three' water coalitions (agricultural users, urban users and environmentalists) and the DWR together. California supported a federal programme (the 1987 Clean Water Act) that enshrined collaboration. In 1991, the state passed a law requiring the California Water Plan to be developed with extensive public involvement from urban, agricultural and environmental representatives (i.e. coalitions) (DWR,

1993). Governor Wilson emphasised his belief in coalition cooperation in an April 1992 speech when he said that all parties would be treated equally with none gaining at the other's detriment (Loh, 1994). Seeds of change were thus planted.

The policy core beliefs of the DWR (1987–2018)

The aftermath of the 1986 floods prompted internal and external anxiety about the DWR's ability to navigate recalcitrant water problems. The DWR's fundamental mission has always been to deliver water to Californians, but they stated that 'evolving environmental policies have introduced considerable uncertainty about much of the State's water supply' (DWR, 1993, 1). Changing policies were forcing new approaches to old problems, and the agency doubted whether it could keep pace with them. It was left to the media to say explicitly that local communities were overly dependent upon imperfect structural measures, shown by the headline 'Rapid rise of Napa River water took local officials by surprise' (Carson et al., 1986). The media also contradicted the DWR's human-centric and construction-oriented policy core beliefs. With the headline 'Storm reminds us we cannot control nature' (Nevada State Journal, 1986), the media admitted candidly what the DWR could not. Taken together, the narrative that emerged after the 1986 floods showed a new level of introspection and increased dissatisfaction with the status quo both within the agency and in California more broadly.

As new laws compelled the DWR to negotiate a course among the various coalitions it served, the agency's historically intractable policy core beliefs broadened to include ecological concerns. During internal committee discussions in 1992, the agricultural coalition argued that infrastructure development was related to national strength and security (Loh, 1994). Usually a potent argument to the agency, this time it fell flat. Instead, the environmental coalition successfully contended that non-structural water management benefits all, as opposed to the state choosing winners (agriculture) and losers (wildlife). In a victory, winter-run Chinook salmon and the Delta smelt became protected under state and federal Endangered Species Acts, and water for fish and wildlife was reallocated from agriculture under the Central Valley Project Improvement Act in 1992.

DWR policy thereby now accommodated ecological demand in addition to human demand, a true shift in policy core beliefs. Moreover, before the release of the 1993 update (DWR, 1993), the DWR stated they were working with the advisory committee to make the 'water plan' a 'more technically accurate and politically balanced document' (Ambruster, 2008, 19). The indication is that past updates were inaccurate and unbalanced. By rejecting the past, the DWR again signals a change in its values. As Democratic Congressman George Miller put it: 'it was a critical turning point in the future of California water policy' (Ellis and Cone, 1993, 2).

Additional legislation proceeding from the 1999 Democratic transition illustrated ecologically focused policy core beliefs. The Poochigian Bill (1999) required the DWR to use an advisory committee in the planning process, and to include the pros and cons of strategies concerning new storage facilities, conservation and recycling, desalination, conjunctive use and water transfers. For the first time, the DWR was required to strategise beyond infrastructure and to create plans for ecological sustainability based on ecological and biological data. Moreover, ecological strategies could not be superficial; the Burton Bill (2000) required the DWR to outline the assumptions that would underlie the plan in a report that was due one year into the planning process. This was a direct response to criticism lobbied from the environmentalist coalition that the underlying assumptions of the 1998 Plan were biased against ecological goals.

The culmination of the consolidation of ecosystem-centric policy core beliefs is shown by legislation regarding integrated water management (IWM) in 2005 by the DWR (DWR, 2005). IWM is a strategic approach, providing multiple benefits to the state's diverse communities and speeding up water projects through broader support (DWR, 2013). DWR started to use IWM with flood management programmes in 2006 under Proposition 64 by gaining $1 billion towards an IWM grant programme. Likewise, Proposition 1E (the Disaster Preparedness and Flood Prevention Bond Act of 2006) gave $4 billion in bonds to repair vulnerable flood structures and protect drinking water supply systems. Further legislation in 2007 and 2008 had the same effect for flood planning in the Central Valley. The idea of IWM is now fully embedded within DWR ideology, as evidenced by mission statements taken from water updates post-2005 (Table 9.2). Flood-related IWM can therefore be understood as the fruition of policy core belief change that took place over the 1993 and 1998 water plan cycles.

Analysis of policy-oriented learning within the DWR

Policy-oriented learning (POL) is a mechanism by which policy change can occur. Only in the context of the Hungarian situation has the contribution of POL to flood management policy change been explicitly examined (Albright, 2011). Though under-applied in real-world examples (Weible et al., 2011), the theory of POL has been defined carefully: it is a long-term process that produces change in thought and behaviour, as well as alters the beliefs of actors in the policy subsystem (Sabatier and Jenkins-Smith, 1999). Sabatier and Jenkins-Smith (1993) contend that POL depends on the amount of conflict among the coalitions, the tractability of the policy problem and, most importantly, whether coalitions can assemble in periodic and active professional forums to learn together. Seen from this perspective, POL is operational: belief change *and* institutional change indicate learning. Moreover, instances of non-learning follow logically from theoretical definitions. That is, non-learning results when the conditions for learning are present à la Sabatier and Jenkins-Smith (1993), but neither belief change nor institutional change takes place.

There is little evidence of POL within the DWR before 1999 because, first, professional forums (i.e. the advisory committees) were contentious and, second, the policy problems were intractable. Successful professional meetings are defined

TABLE 9.2 A summary of stated DWR missions, key actors involved in water plan updates, and the corresponding policy core beliefs (taken from the DWR water plans 1987–2018)

Update year	DWR mission	Key actors	Policy core beliefs
1987	Take a broad view of water issues	Republican-controlled state government	Pro-development
1993	Attempt to balance agric./urban/environmental interests; planning process included public participation	Professional forum: an advisory committee of Big Three coalitions	Balance between ecologically focused and development
1998	Pivot from a technical exercise towards a multi-stakeholder consensus-seeking planning process	New DWR directors Jonas Minton and Lestor Snow	Ecologically focused
2005	Collaborative review process; beginning of IWM	DWR director and staff	Sustainability focused

as events where opposing coalitions reach consensus and negotiate shared solutions (Sabatier and Jenkins-Smith, 1993). This was not occurring before 1999. Some stakeholders on the advisory committee leading up to the 1998 update noted that DWR staff were 'often unwilling to devote time to discussing some of the broader philosophical issues of the water plan updates' approach' (Ambruster, 2008, 37), heightening tensions. Federal actor Felicia Marcus described the forums as a war zone: 'We are trying to end the gridlock on water policy as opposed to constantly firing missiles at each other' (Ellis and Cone, 1993, 1).

In addition to the hostile atmosphere, the water policy problems were becoming increasingly challenging, which halted POL. For instance, new protections were needed to restore the Delta and San Francisco Bay. By any standard, these restoration projects would be difficult. However, the situation became particularly difficult when Republican Governor Pete Wilson – who was committed to urban, industrial and agricultural interests over environmental protections – went so far as to dispute data provided from the federal government concerning the restoration. Wilson said: 'The federal agencies continue to misrepresent the water costs of their actions' (Ellis and Cone, 1993, 1). Wilson's positions further strangled coalition meetings when he imposed a strict deadline on the DWR to finish the update before his term ended in January 1999. This may have stymied any good faith attempts DWR staff had to work with the coalitions and thus stalled the learning process.

Yet 1999 was a turning point. The transition to a Democratic governor in 1999 brought a progressive management philosophy and new leadership to the DWR. Incoming governor Davis hired new DWR managers, and many DWR staff were replaced, moved or retired. Jonas Minton, who successfully directed the Sacramento Area Water Forum consensus-seeking effort, was made deputy director.

Minton introduced the idea of collaborative planning to the DWR and showed he was keen to disseminate this idea. For example, Minton began workshops with the public and DWR staff, which had never occurred before. As DWR staff interacted with those outside of their agency, they gained new insight into the broader context of water policy, a key part of POL. Then Lester Snow, the former executive director of the collaborative CALFED Bay Delta programme, became the DWR director. Snow had direct experience finding common ground among representatives of state and federal operators, regulatory agencies and stakeholders of the three big water interest groups (Public Policy Institute of California, 2018). He put together an advisory committee that was made up of 65 people with diverse interests, and DWR staff had to cope with a new collaborative culture. Thus the 1999 administration generated the conditions necessary for POL to occur.

With a new director committed to collaboration, major institutional changes allowed POL to continue through the lead up to the 2005 Update. This time POL was taking place between agricultural and environmentally oriented stakeholders during meetings, as evidenced by interviews conducted during the discussion phase (Ambruster, 2008). For example, pro-environment staff originally urged the practice of targeted crop stress as a water-saving technique, but after agricultural staff gave presentations that cast doubt on the strategy's effectiveness, they withdrew support of the idea and it was dropped from the draft (Ambruster, 2008).

DWR staff continued to learn. Formally, the DWR had total authority over the final contents of their updates, but in practice the agency now accepted nearly all the consensus decisions put forward by the advisory committee. This represented a significant change from the approach a decade earlier. When published, the plan called for renewed emphasis on urban and farm conservation, increased groundwater storage, urban water recycling and farm runoff desalination. Collaboration had succeeded. The press took note of just how far the water policy arena had come; an editorial in the Los Angeles Times was entitled 'A Shocking Water Noncrisis' (Los Angeles Times, 2005). These examples all provide evidence for the diffusion of a collaborative culture and POL, which spread from agency leaders beginning in 1999.

Conclusions

The Advocacy Coalition Framework (ACF) is a powerful lens through which to examine a policy process under changing political, social and hydrological circumstances.

This AFC analysis of policy belief change in California demonstrates how flood risk management came to sit within broader paradigms of integrated water management rather than being seen as a separate and engineering dominated domain. Although the ACF does not exhaust all the ways to describe the policy process, this example illustrates that elements of the ACF can be fruitfully applied to describe how the construction-orientated mentality within the California Department of Water Resources (DWR) gave way to the outlook of integrated resource management over a period of two decades. This was as a result of an increasingly influential – and publicly supported – broad coalition of environmental interests gradually came to exert its influence.

One of the ACF's key strengths lies in its acknowledgment of individual psychological factors: individuals have deeply held beliefs that shape their fundamental policy positions (Weible and Sabatier, 2006), which other policy frameworks neglect.

Despite the limitation of requiring a long-term perspective of change (>10 years), the concept of POL and core beliefs provide pathways to identify the theoretical, practical, and operational conditions that must be met for policy change to occur. This chapter contributes to the literature that has often omitted a discussion of important psychological and social processes involved in the development of modern flood risk management.

References

Albright, E. (2011). Policy Change and Learning in Response to Extreme Flood Events in Hungary: An Advocacy Coalition Approach. *Policy Studies Journal*, 39(3), 485–511.

Ambruster, A. (2008). *Collaborative Versus Technocratic Policymaking: California's Statewide Water Plan*. Sacramento: Center for Collaborative Policy.

Biswas, A. (2004). Integrated Water Resources Management: A Reassessment. *Water International*, 29(2), 248–256.

California Department of Water Resources (1957). California's Flood Future: Public Draft Report: History of California Flooding. Bulletin 3, Sacramento, CA.

Carson, L., Ernst, D. and Courtney, K. (1986). Rapid Rise of Napa River Water Took Local Officials by Surprise. Napa Valley Register. [Online] Available at: https://napavalleyregister.com/news/local/article_021c4834-3af8-11e0-b171-001cc4c002e0.html [Accessed 20 Feb. 2018].

DWR (California Department of Water Resources) (1966). Bulletin 160–66. DWR, Sacramento, CA.

DWR (California Department of Water Resources) (1970). Bulletin 160–70. DWR, Sacramento, CA.

DWR (California Department of Water Resources) (1974). Bulletin 160–74. DWR, Sacramento, CA.

DWR (California Department of Water Resources) (1983). Bulletin 160–83. DWR, Sacramento, CA.

DWR (California Department of Water Resources) (1986). The Floods of February. DWR, Sacramento, CA.

DWR (California Department of Water Resources) (1987). Bulletin 160–87. DWR, Sacramento, CA.

DWR (California Department of Water Resources) (1993). Bulletin 160–93. Sacramento, CA.

DWR (California Department of Water Resources) (2005). Bulletin 160–05. DWR, Sacramento, CA.

DWR (California Department of Water Resources) (2010). Central Valley Flood Management Planning Program: State Plan of Flood Control Descriptive Document. DWR, Sacramento, CA.

DWR (California Department of Water Resources) (2013). California's Flood Future: Recommendations for Managing the State's Flood Risk. DWR, Sacramento, CA.

DWR (California Department of Water Resources) (2018). Bulletin 160–18. DWR, Sacramento, CA.

Ellis, V. and Cone, M. (1993). Wilson to Cooperate with U.S. Water Plan. *Los Angeles Times*. [Online] Available at: http://articles.latimes.com/1993-12-16/news/mn-2406_1_delta-water [Accessed 20 Feb. 2018].

Ferguson, L. (2017). Why the City of Napa Did Not Flood in 2017: How an Environmentally Designed Flood Control Project Worked. Regional Water Quality Control Board, San Francisco.

Jenkins-Smith, H. and Sabatier, P. (1993). The Dynamics of Policy-Oriented Learning. In P. Sabatier and H. Jenkins-Smith (eds), *Policy Change and Learning: An Advocacy Coalition Approach*. Boulder, CO: Westview Press.

Jenkins-Smith, H. and Sabatier, P. (1994). Evaluating the Advocacy Coalition Framework. *Journal of Public Policy*, 14(2), 175–203.

Kelley, R. (1989). *Battling the Inland Sea: Floods, Public Policy, and the Sacramento Valley, 1850–1986.* Berkeley: University of California Press.

Los Angeles Times (2005). A Shocking Water Noncrisis. *Los Angeles Times*. [Online] Available at: http://articles.latimes.com/2005/jun/15/opinion/ed-water15 [Accessed 25 Feb. 2018].

Loh, P. (1994). (De)Constructing the California Water Plan: Science, Politics, and Sustainability. Unpublished master's thesis, Energy and Resources Group at the University of California, Berkeley.

Meijerink, S. (2005). Understanding Policy Stability and Change: The Interplay of Advocacy Coalitions and Epistemic Communities, Windows of Opportunity, and Dutch Coastal Flooding Policy 1945–2003. *Journal of European Public Policy*, 12(6), 1060–1077.

Nevada State Journal (1986). Storm Reminds Us We Cannot Control Nature. *Nevada State Journal*. [Online] Available at: http://thestormking.com/Sierra_Stories/1986_Flood_Disaster_/Flood_Editorial/flood_editorial.html. [Accessed 25 Feb. 2018].

Public Policy Institute of California (2018). Lester Snow: Bio. [Online] Available at: www.ppic.org/person/lester-snow/ [Accessed 1 Mar. 2018].

Sabatier, P. (1987). Knowledge, Policy-Oriented Learning, and Policy Change: An Advocacy Coalition Framework. *Knowledge: Creation, Diffusion, Utilization*, 8(4), 649–692.

Sabatier, P. (1993). Policy Change over a Decade or More. In P. Sabatier and H. Jenkins-Smith (eds), *Policy Change and Learning: An Advocacy Coalition Approach* (13–39). Boulder, CO: Westview Press.

Sabatier, P. (1999). *Theories of the Policy Process: Theoretical Lenses on Public Policy.* Boulder, CO: Westview Press.

Sabatier, P. A. and Jenkins-Smith, H. (1993). *Policy Change and Learning: An Advocacy Coalition Approach.* Boulder, CO: Westview Press.

Sabatier, P. and Jenkins-Smith, H. (1999). The Advocacy Coalition Framework: An Assessment. In P. Sabatier (ed.), *Theories of the Policy Process*, 1st edition. Boulder, CO, Westview Press, 117–166.

Schlager, E. (2007). A Comparison of Frameworks, Theories, and Models of Policy Processes. In P. Sabatier (ed.), *Theories of the Policy Process*, Boulder, CO: Westview Press, 293–319.

Taylor, M. (2017). Managing Floods in California. Legislative Analyst's Office, Sacramento, CA.

Weible, C. and Sabatier, P. (2006). A Guide to the Advocacy Coalition Framework. In F. Fischer, G. Miller and M. Sidney (eds), *Handbook of Public Policy Analysis: Theory, Politics, and Methods*. New York: CRC Press, 123–136.

Weible, C., Sabatier, P., Jenkins-Smith, H., Nohrstedt, D., Henry, A. and de Leon, P. (2011). A Quarter Century of the Advocacy Coalition Framework: An Introduction to the Special Issue. *Policy Studies Journal*, 39(3), 349–360.

Yuba County Water Agency (2012). History of Flooding and Flood Control. [Online] Available at: www.bepreparedyuba.org/pages/prepare/history.aspx [Accessed 2 Mar. 2018].

10

THE CHALLENGES OF FLOOD WARNING SYSTEMS IN THE DEVELOPING WORLD

Mahala McLindin

Introduction

The impacts of flooding

Flooding is a 'profound cause of underdevelopment' (Briscoe, 2009, 19). The impacts of fluvial floods are skewed towards developing countries, which experience the most severe floods with greater losses of life, possessions and livelihoods than floods in developed countries (ADB, 2003; Bakker, 2009). Some 80 per cent of people exposed to river flood risk are located within just 15 developing countries, with highest exposure in India, Bangladesh, China and Vietnam (WRI, 2015). Agricultural vulnerability in these countries can cause further disproportionate impacts that, with repeated flood events, cumulatively engrain inter-generational poverty (Webster and Jian, 2011).

Developing countries often lack the economic resources, institutions and infrastructure to minimise these flood impacts and foster resilience. Conversely, however, floods can be crucial to livelihoods and economic growth in the same locations, for example in Bangladesh and Cambodia. Flood risk management interventions must therefore tread a fine line between reducing impacts and retaining benefits, all within the context of limited financial resources.

Flood early warning systems

In the developing world, flood early warning systems (FEWS) are seen as one of the best ways of balancing the impacts and benefits of floods. Warnings with sufficient lead times can reduce fatalities, limit economic loss and allow for timely agricultural adaptation (Hallegatte et al., 2016). Recognising this, a global imperative to enhance early warning systems in developing countries was instigated by the

UN Hyogo Framework for Disaster Reduction (2005–2015) and continues under the Sendai Framework (2015–2030) and Sustainable Development Goals.

Despite such impetus, FEWS remain underdeveloped. In the 2010 Indus floods in Pakistan, for example, an effective FEWS could have provided six days' notice of approaching floodwaters. The floods killed 2,000 people, cost US$40 billion and directly impacted 18 million people (BBC, 2010). Where FEWS do exist, research frequently demonstrates that lead times are inadequate, warnings fail to reach at-risk communities and response behaviour is deficient. Achieving an effective FEWS is therefore universally challenging. Even countries like Australia and the UK continue to face well-researched failures and inefficiencies within their FEWS (Parker, 2017). Despite the disproportionate impacts of flooding in developing countries and demonstrable benefits of effective early warning, the literature demonstrates inadequate research into the challenges undermining effective FEWS there.

The objective of this chapter

Through a review of literature and case studies, this chapter explores the key challenges for effective FEWS in developing countries, aiming to understand recent approaches to overcome them. It argues that fundamental challenges occur where there is a difference between who possesses the necessary *knowledge* to make a flood warning system effective and who *controls* that system; we therefore describe disjunctures in the location of knowledge and control.

The analysis is framed by the concept of an early warning system 'chain' (Parker, 2017) where the strength of the whole chain is no greater than that of its weakest link. For each component or link in that chain we explore who is in control, who possesses the required knowledge, and the ensuing challenges when that knowledge and that control diverge. By 'control' here we mean the authority to design, administer, monitor, review and take corrective action related to the FEWS. By 'knowledge' we mean the skills, awareness and information acquired through education or experience.

The early warning system 'chain'

The FEWS chain is a cumulative set of actions that aims to achieve appropriate preparedness and loss minimisation measures when a flood occurs. Parker (2003) outlines five key components of the chain: detection, forecasting, warning, response and learning (see also chapter 11, Figure 11.1).

Each component or link of the chain requires differing and specific 'formal' and 'vernacular' knowledge for its effective design, development and administration. The initial technocratic components of flood detection and forecasting require formal knowledge of hydrology and meteorology and the spatial and computational sciences. The dissemination of flood warnings requires formal knowledge of communication technologies. The development and implementation of response strategies requires knowledge of spatial and communication systems alongside community insights and infrastructure planning and operation.

'Vernacular knowledge' is equally required, developed through experience within the local context. For instance, the warning component requires experiential and social knowledge of appropriate technology and social structures, as well as organisational knowledge to administer the warning through and across institutional structures. The response component requires similar social knowledge to understand how individual and community factors influence behaviour.

As we move from the first to the last component of the chain, there is a shift in the dominant type of knowledge required from formal scientific knowledge to local vernacular knowledge. There is a corresponding shift in who possesses the requisite knowledge to manage the component effectively, from centralised organisations to local communities. Hence, a shift in the *location* of knowledge occurs to match the shift in the *type* of knowledge required.

Each component of the chain requires different organisations to control its design and operation. For example, control of detection and forecasting requires an authority with access to hydrometeorological infrastructure and technical capacity. Control of warning dissemination requires an authority capable of managing communications infrastructure that penetrates vulnerable communities. Response requires control by authorities with local facilities for emergency action.

This chapter develops Parker and Handmer's (1998) idea to argue that an effective flood warning chain in the developing world needs the transitioning 'location of knowledge' to be matched with an equivalent shift in the location of the controlling agent or authority. For example, the components of the chain requiring formal scientific knowledge are best controlled at a central level in each country, while the components requiring local vernacular knowledge are best controlled at the community level. Therefore, a shift in the 'location of control' is required (Figure 10.1).

In practice, aligning the location of knowledge and control presents a complex interdisciplinary challenge, requiring coordination both vertically down government hierarchies and horizontally across disciplines and sectors of industry and society. For success, effective functioning of each component needs to be

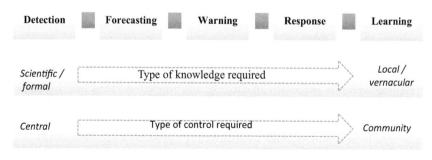

FIGURE 10.1 The 'chain' in flood early warning systems (adapted from Parker, 2003).

combined with effective integration along the whole chain. Natural disasters will expose cumulative process deficiencies in the chain, and it is the increased like-lihood of unsuccessful FEWS coordination in developing countries that leads to the increased impacts from natural disasters there (World Bank, 2010).

Detection and forecasting

Detection and forecasting are the first, highly technical, components of the early warning system chain, requiring expert scientific knowledge. In FEWS in devel-oped countries such as Australia and the UK, effective knowledge comes through well-funded, national technical organisations. Control is correspondingly located within national organisations with access to global data and advanced technology. Developing country forecasting is also controlled at a national level, but challenges occur when these organisations do not possess adequate technical knowledge and skills for effective detection and forecasting. Here such knowledge is impeded by resource deficiencies and, in the case of transboundary basins, inadequate regional cooperation. The result is forecasts with inadequate accuracy or lead times to pro-vide effective evacuation focused warnings, let alone livelihood protection mea-sures and thus economic resilience.

Resource challenges

Knowledge to achieve accurate and timely flood forecasting is frequently impeded throughout the developing world by critically deficient infrastructure within national hydrometeorological services (WMO, 2008). In more than a hundred countries, sustained underfunding has undermined the ability of these hydro-meteorological services to effectively reduce disaster risk (Rogers and Tsirkunov, 2013). This is particularly pertinent in Africa, where 44 out of 54 countries reportedly have inadequate meteorological services (World Bank, 2016). Under-funding limits the installation of infrastructure, including observational networks, computing resources and communications facilities. National agencies then cannot access and process hydrological data and global weather information, knowledge necessary for generating local flood predictions. Senegal, for example, faces severe recurrent flooding, yet the national meteorological agency has stated that forecast-ing of major floods in 2016 was impossible due to inadequate rainfall gauging equipment (Niang, 2016).

Where infrastructure does exist, underfunded maintenance has led to the dete-rioration or abandonment of equipment. This challenge is particularly relevant in developing countries where donor funding may support infrastructure installation but not ongoing operation or maintenance costs. In Sub-Saharan Africa, for example, up to three quarters of installed weather stations no longer function, and similar trends are reported across Central Asia (Rogers and Tsirkunov, 2013).

Knowledge within national agencies in developing countries is also undermined by inadequate staff technical capacity. Inappropriate training and insufficient technical skills within national agencies throughout least developed and developing countries limit the knowledge required to develop and interpret flood forecasts (Theimig et al., 2011; WMO 2008).

Transboundary challenges

The disproportionate severity and impacts of transboundary floods (Bakker, 2009) makes effective forecasting crucial for flood-prone developing nations located at the downstream end of international river basins, such as Pakistan, Senegal, Bangladesh and Vietnam. Here, however, a disjuncture occurs between the location of control at national level and the required hydrometeorological knowledge located at the broader basin level. To overcome this, meaningful regional cooperation is required between riparian nations through either data sharing to equip the national controlling organisation with the requisite knowledge or the development of regional forecasting organisations that shift control from the national to the basin level. However, poor regional coordination is endemic to most flood-prone developing nations and severely impedes effective flood detection and forecasting (Hossain, 2007).

The presence of basin level organisations correlates with lower financial flood damages, deaths and displacement tolls (Bakker, 2009). However, only 78 out of 279 international river basins have transboundary water institutions and – as highlighted by the Nile, Indus, Zambezi and Amazon – very few of these countries that suffer from recurring transboundary floods have institutions incorporating flood issues. Fewer again include direct platforms for data exchange (Gerlak et al., 2011). Least developed and developing countries, in particular, have identified poor regional data exchange as a key barrier to effective flood detection and forecasting (WMO, 2008). Bakker and Duncan (2017) highlight particular vulnerabilities in Africa and Asia, where increasing hydrological hazards and rising populations intersect in transboundary basins without regional flood management.

The Mekong River Commission illustrates some more effective regional flood forecasting cooperation. The Commission gauges the main stem of the Mekong and collates tributary data provided by Laos, Thailand, Vietnam and Cambodia. The Commission shares data between member countries and generates forecasts and warnings for integration within existing national systems. But the Commission also presents a common challenge for regional systems in developing contexts. For successful international collaboration in transboundary basins, appropriate national institutional structures, regulatory systems and infrastructure compatible with other riparian states are first required (and some sovereignty thereby sacrificed); these systems are often lacking in developing countries. Research on the Mekong highlights national impediments that prevent effective movement of knowledge to the regional level. For example, data sharing is hampered by conflicting data security policies in Vietnam, by limited observational infrastructure in Laos, and by a lack of institutional support in Thailand due to a perception there of limited accruing benefits (Plengsaeng et al., 2014; Thu and When, 2016).

Regional cooperation requires the challenge of synchronising all aspects of riparian countries' policies, legislation and procedures (Bakker and Duncan, 2017). Such cooperation can be undermined by riparian power dynamics, fuelled by economic competition for basin resources or political mistrust between basin members (Bakker, 2009). For example, data sharing between China, India and Pakistan for the Indus River is seen as a 'lofty dream' due to conflicts arising from the Indus Water Treaty (Zia and Wagner, 2015, 189). In the Ganges basin, conflict over water management and infrastructure development, alongside prevailing political dynamics, also impedes cooperative river management. There is no multilateral agreement between China, India, Nepal and Bangladesh to share real-time hydrological data. For downstream Bangladesh, access to all upstream data would allow flood forecasts with 14 days' lead-time. Instead, restricted to data from gauges in Bangladesh alone, the national forecasting agency can only provide forecasts with a lead time of two days (Hossain et al., 2007).

Warning and dissemination

The warning component of the chain marks a transition point in the location of knowledge. In the previous component of the chain, knowledge was only located at the macro (inter)national level with dependency upon technical expertise. Here, technical knowledge of dissemination technologies requires assimilation with local flood-related knowledge for effective warning penetration.

A similar transition in the location of control also occurs at this point in the chain. Warning dissemination is initially controlled by a national agency following detection and forecasting, but effective FEWS require the warning to pass successfully to local-level authorities, which can then target warnings to communities known by them to be vulnerable.

A disjuncture in the locations of knowledge and control arises here if centralised management does not transition towards community control, in line with the shifting knowledge requirements. Overcoming this point of 'discontinuity in communication' is a global challenge (Parker and Handmer, 1998, 47). These issues have been explored in developed countries to leverage significant investments in the preceding sections of the chain (Parker, 2003), and while these learnings are relevant, specific contextual issues are evident in developing countries where control and knowledge diverge.

In particular, challenges occur where institutional structures through which the warnings transfer from disseminator to recipient cause impediments. Where controlling authorities do not possess adequate knowledge of the contextual factors that influence appropriate technology and social vulnerability, this can lead to technological and social exclusion from warning systems. These bureaucratic inefficiencies have been cited as a key reason for delays in the warning processes of developing countries (Keoduangsine and Goodwin, 2012).

Institutional impediments

Well-structured institutional frameworks are required to match the location of flood management control with the transitioning location of knowledge. This presents a challenge in developing countries, where controlling institutions are often weakened by ineffective frameworks, unstable politics and frequent change (Parker and Handmer, 1998). Here, national government institutions are generally the sole issuers of official warnings. Reliance on such national-level processes (see chapter 11) limits the effectiveness of warnings to successfully transition down through institutional hierarchies to alert threatened communities (WMO, 2008), as national agencies often lack local knowledge of flood impacts, appropriate technology and community traits. Despite effective internal government communication, warnings reportedly fail to reach the public in Bangladesh for these reasons (Mahmud et al., 2012).

Failings in institutional coordination within and across agencies also impede the alignment of knowledge and control. In research on flood-prone communities in Bangladesh, Rahman et al. (2012) highlight poor coordination between different levels of response organisations that impedes FEWS effectiveness. A review of the 2014 Karnali River flood in Nepal that killed 222 people similarly highlighted poor leadership, coordination and relationships within and between response organisations and sectors (MacClune et al., 2015). Further issues included the absence of monitoring and evaluation processes, which prevented institutional learning from past events such as the 2008 Koshi floods, or the transfer of experiential knowledge to controlling authorities.

Technological exclusion

Technology also plays a crucial role in warning dissemination, with appropriate selection a key challenge. Social media are increasingly complementing traditional technologies in providing warning notifications. However, challenges of access, acceptance and uptake can limit effectiveness (see chapter 16). For example, disadvantaged groups can be excluded from receiving warnings sent to mobile phones or social media (Parker, 2017). Poor technology selection demonstrates a failure of the controlling authority to understand individual and community needs, the requisite knowledge of which lies with the warning recipient.

In developing countries these global challenges are most acute in locations with limited financial resources, intermittent power supply, low literacy and education levels, and the potential for theft and vandalism of warning infrastructure (Basha and Rus, 2007). In the mountainous regions of northern Vietnam, for instance, the national early warning system failed to reach ethnic minority groups because the centralised communication system was made ineffective through power shortages, equipment failures and linguistic barriers (PAMC, 2012).

However, mobile telephone technology is now ubiquitous in developing countries such as Kenya and Bangladesh. Its application to flood warnings is increasingly evident, but is still seen as underutilised as an effective warning tool (Cumiskey, 2015). In a study of a mobile-enabled FEWS in Bangladesh, recipients demonstrated a preference for SMS warnings, particularly using voice SMS or symbols to overcome language and literacy issues (Cumiskey, 2016). In Bhutan, SMS warnings were specifically preferred by women, who were more likely to be undertaking domestic or agricultural work beyond the reach of sirens or loudspeakers. This example reinforces the need to understand local context alongside recipient preferences, because frequent mobile network failure during disasters in Bhutan limits their suitability within FEWS without backup methods (Shrestha et al., 2016).

Social exclusion

Geographic, social and political exclusion of communities from warnings can also result from centrally controlled systems that fail to incorporate community knowledge of social vulnerability. Effective warning penetration is highly pertinent to the developing world, where warnings often fail to reach the most vulnerable. India's national flood warning system, for example, does not reach the most vulnerable in Bahraich, Utta Pradesh – the most flood-affected district in the most flood-prone state of India (PAMC, 2012). In these contexts, development and settlement patterns put marginalised groups at greater risk through occupation of high-risk areas that are outside the remit of centralised warning systems and beyond the capabilities of traditional communication techniques (Basha and Rus, 2007)

Marginal groups also carry little influence over political decisions and are unlikely to contribute their local knowledge to the development of warnings, particular in centrally structured systems. Their warning needs may therefore be overlooked. A review of a FEWS in Bhutan found that 40 per cent of the target population who were illiterate were unaware of hazard zones and escape routes because information on warning and response plans was distributed using only written material (Shrestha et al., 2016).

Political exclusion may also occur where conflict exists between the state and marginal communities. Given the conflict between the Pakistani and Afghani governments and Pushtan/Taliban fighters in the upper Indus basin, national flood warnings are unlikely to target these groups while the government is said to be actively engaging in misinformation (Zia, 2015). Similarly, in Myanmar, political conflict has historically impeded engagement between government and flood-prone Rohinhya communities in Rakhine state (PAMC, 2012).

Response

Response behaviour is influenced by highly contextual factors relating to individual flood experience, psychology and the local social context shaping cultural beliefs and behaviours. Understanding how these factors affect human response to flood warnings therefore requires detailed knowledge of the local context. Where authorities in control of the response component of the chain fail to understand and incorporate local knowledge into response design, planning and implementation activities, ineffective responses are likely at the local level, leading to failure of the whole system.

Mileti (1995) theorised that improved flood warning response occurs when sender and receiver characteristics are alike. This could be adapted to argue that benefits are achieved by aligning the *knowledge* of sender and receiver. Disparities of knowledge between disseminators and receivers/responders are addressed in well-resourced systems as in Australia, where response plans are collaboratively developed between local authorities and community groups. In contrast, the control of Bangladeshi flood response planning and activities has historically been managed by the national disaster agency without community involvement. The disparate characteristics and knowledge of sender and receiver here reduces the controlling authority's understanding of how individuals and communities will respond. The need of controlling authorities to possess and incorporate local knowledge into FEWS is illustrated by looking at the influence of flood experience, psychology and social context.

The influence of experience

Experience of previous flood events is widely acknowledged to improve responsiveness to warnings. In northern Ghana, recurrent flooding of the White Volta River has taught communities to understand natural flood indicators, including inland migration of hippo populations prior to flood events. Excluded from official warning systems due to isolation and poverty, the hippos provide the community with valuable insight to initiate flood response actions (Ngwese et al., 2018).

However, the effectiveness of experience-based response may be being eroded as climate, development and migration change flood predictability and impact knowledge accumulation. For example, Paul and Routray (2010) noted that indigenous forecasting and coping practices are effective in normal floods in Bangladesh but have limited benefit in extreme floods. Past experience of extreme or recurrent events may also generate the desire to respond, but the losses incurred may make action unaffordable. This is particularly relevant in developing countries, where floods have greater economic impact relative to income, and repeated flooding can damage livelihoods and resilience. Following major Brahmaputra flooding in 2007, for example, a survey of rural floodplain communities demonstrated that flood response decisions were heavily influenced by people's ability to bear the associated costs (Sha et al., 2012).

For those without prior flood experience, a lack of understanding of how to assess or respond to risk can prevent action. People assess risk based on past available or imagined experiences – termed the 'availability heuristic' – which can underestimate potential consequences and the required response (Parker, 2017). This is particularly relevant in locations with transient, migrant populations or infrequent flooding where inexperience is coupled with some apathy. Plate (2007) highlights this challenge for communities living on the edge of the Mekong floodplain, where preparedness is forgotten because floods recur only on decadal timescales.

Psychological influences

Psychological factors, including common reactions of disbelief, denial or distrust, can lead to constrained response to flood warnings. Disbelief is common where warnings are received from distant, bureaucratic or unreliable organisations. During cyclone-induced flooding that killed more than 138,000 people in Bangladesh in 1991, disbelief of warnings was a key reason for inaction by up to 70 per cent of the affected population. Disbelief, however, can prompt people to seek new information from trusted sources and social networks in the common warning confirmation process (Mileti, 1995). But response may thereby be delayed. This process of belief confirmation may be prolonged in developing countries where technology, transport and geographical constraints can impede efficient communication.

Trust in flood authorities also influences response behaviour. A survey of flood-prone urban settlements in Mumbai found that warnings issued by trusted government sources enhanced acceptance and adoption of flood mitigation action (Samaddar et al., 2012). Conversely, technical and bureaucratic issues at a macro level influence this trust in information – and in turn response – at a micro level. Distrust in the provision of government support or structural defences, for example, can impede action. Residents of floodplains in China reportedly place minimal trust in government intervention, instead adopting self-activated coping strategies (Lamond and Proverbs, 2009). In Indonesia, residents of informal riverbank settlements did not evacuate after receiving warnings of flooding in 2010. They distrusted the government due to its inaction in solving flooding issues, and suspected the flood warnings promoting evacuation were part of a political agenda for slum clearance (van Voorst, 2014).

The influence of social context

Human behaviour is strongly influenced by the social context that develops and sustains cultural beliefs, knowledge and attitudes. With floods, people's perception of risk influences how and whether they will respond, and this perception is shaped by local contextual factors.

In developing countries, livelihood decisions affect all such behaviours. Reliance on continuing agricultural practices, protecting property from looting and concerns surrounding vacating properties with no security of title (World Bank, 2010) may all contribute to delayed evacuation response to warnings. Religious belief can be important. Passive response to flood warnings as a result of religious fatalism has been reported in communities on the Lempa River, El Salvador (Schipper,

2010) and the Volta River, Ghana (Bempah and Øyhus, 2017). Additionally, response behaviours are derived from cultural practices of ethnic groups: a study of flood-prone communities in Nigeria found a statistically significant difference in flood response between different ethnic groups (Boamah et al., 2015).

Wider gender inequity in developing countries also generates cultural norms that strongly influence female response to warnings. Cultural restrictions around physical behaviour and dress inhibit women's swimming and running capabilities during floods. Less formal education and heavy workloads also constrain women's access to flood-related information. For example, a review of a FEWS in Punatsangchhue River, Bhutan, found that lower attendance at training sessions meant the majority of local women were unaware of safety routes and evacuation sites (Shrestha et al., 2016). Stifled decision-making powers, fear of gender-based violence, reluctance to face public exposure, and child-care responsibilities further contributed to delayed action by women. However, upon hearing an FEWS siren in Bhutan, male focus group participants stated they would immediately assemble family members and neighbours before moving to safety (Shrestha et al., 2016). Gender vulnerability in flood response was made strikingly evident in the 1991 Bangladesh floods described above, where 90 per cent of fatalities were female. Men reportedly received warnings in public spaces, while women perished as result of cultural norms that restricted women from leaving their homes alone (Paul et al., 2011).

Addressing challenges in developing countries

There are many challenges within each component of the flood early warning chain. There are, however, many examples of systems that address some of these challenges by shifting from a top-down, centralised approached toward solutions that better delegate control to those with the appropriate knowledge and understanding. These examples embrace two approaches: official community-led systems and collaborative systems.

Official community-led systems

Community-led flood early warning systems provide a pragmatic option for developing countries. Those with experience of catastrophic loss are most adept at developing adaptation strategies (Webster and Jian, 2011), and community involvement harnesses this knowledge to overcome challenges in the latter part of the chain, where inefficiencies flow from a disjuncture in knowledge and control.

To overcome the failings in centralised transboundary flood management in Southern Asia, for example, a community FEWS was piloted on the Ratu River linking upstream Nepali and downstream Indian villages. Water-level sensors transmit information to Nepali caretakers, who disseminate warnings downstream using SMS, sirens or flags. In the 2017 floods eight hours warning for evacuation resulted, and the arrangement has since been replicated in Afghanistan, India and Pakistan (Molden et al., 2017).

Technology issues can be addressed by involving communities in system design. In the Ratu system, simple low-cost technology allowed for local manufacturing, repair and maintenance. A community FEWS near the borders of India and Nepal demonstrated the pragmatic adoption of available technology by using the border police's robust radio systems for backup warnings (PAMC, 2012).

Addressing challenges of exclusion and non-response, communities in the Karnali River basin in Nepal contributed to hazard mapping and identifying evacuation routes, although with manual observations only two-hour warnings can be provided. Systems reliant on volunteer support, with limited infrastructure and competing livelihood priorities can be weak. In the failure of a community system on the Babai River, Nepal, with 20 fatalities (MacClune et al., 2015), community responses were hampered by many trained volunteers not being contactable at night or having left the region.

By shifting control to the local level, community systems can lack the formal knowledge required for the technical warning chain components. They also rely on strong local leadership and sustained local voluntary support, particularly challenging in low-income communities where immediate livelihoods take priority. Such community systems are perhaps only appropriate for evacuation-focused warnings in small river basins with small-scale floods (Basha and Rus, 2007); major flood catastrophes may be unaffected.

Collaborative systems

Collaborative approaches between international agencies, government and communities are increasing in the developing world. These projects use the formal knowledge and resource capacity of the international community to improve the early technical stages of the forecasting and warning chain, then move to community-led approaches in the later components, usefully aligning control with the most appropriate knowledge locations.

For example, an initiative by the World Meteorological Organisation (WMO) and the US Hydrological Research Centre undertakes satellite-based flood forecasting and provides technical assistance to localise and integrate these into national warning systems. Its application to the Kamala River, Nepal, has strengthened national technical skills and flood response initiatives at municipal and community levels. In Niger, WMO are again training forecasters and improving infrastructure at the national meteorological service, allowing access to flood predictions from specialised regional and global centres (WMO, 2017). International technical support is also bringing wider regional cooperation to the Ganges and Indus, through the HKH-HYCOS programme. International organisations are supporting national institutions to improve infrastructure, data sharing, forecasting capacity and communication to end users (ICIMOD, 2018).

Finally, international organisations have supported upscaling of community systems to achieve multi-level integration within centralised systems. The community gauges of the Karnali system are progressively being incorporated into the national

Nepali network, linking national warnings to community committees. Sixteen community FEWS in the Philippines have been developed by the German government, and these have then been integrated up through local and national government systems (Perez et al., 2007).

Through aligning knowledge and control, these cases all exemplify hybrid methods that seek to address technical and resourcing challenges while incorporating localised approaches. Zia and Wagner (2015), however, argue that this focus remains skewed towards technocratic improvements without adequate emphasis on end users: the chain is unbalanced. Even where balance is possible, the significant drawback of both community and collaborative systems can be the potentially unsustainable reliance on external assistance.

Conclusions

Many barriers impede successful FEWS in developing countries. By examining the location of knowledge and control through the detection, forecasting, warning and response components of the FEWS chain, we argue that fundamental challenges occur where knowledge and control diverge. The final 'learning' component of the chain, while not explicitly explored here (see chapter 11), can further support this argument. Few FEWS in developing countries include formal evaluations that can transfer local experiential knowledge up to enhance the institutional knowledge of controlling authorities, thus continuing to promote this disjuncture of knowledge and control.

Over the past few decades, however, increasingly collaborative solutions have been demonstrated to overcome some of these challenges. More recently, the publication of WMO's multi-hazard early warning systems checklist (WMO, 2018) has encouraged a new rhetoric of 'multi-level' solutions, in line with our principal argument here.

But there is no simple way of overcoming all the challenges of FEWS within developing countries. However, their design may be improved by understanding and accounting for the highly contextual challenges outlined here of institutional capacity, technical skills, appropriate technology, social exclusion, experience, psychology and culture. Fundamental to this is developing a system that matches the transitioning location of knowledge along the FEWS chain with an equivalent multi-level shift in the location of the controlling authority.

References

ADB (Asian Development Bank) (2003). Floods and the poor: Reducing the vulnerability of the poor to the negative impacts of floods, ADB, Manila.

Bakker, H. (2009). Transboundary river floods and institutional capacity, *Journal of the American Water Resources Association*, 45(3), 553–566.

Bakker, H. and Duncan, J. (2017). Future bottlenecks in international river basins: Where transboundary institutions, population growth and hydrological variability intersect, *Water International*, 41(4), 400–424.

Basha, E. and Rus, D. (2007). Design of early warning flood detection systems for developing countries. In: *Proceedings: International Conference on Information and Communication Technologies and Development*, IEEE, Bangalore.

BBC (2010). Pakistan floods 'hit 14m people'. Accessed 5. 8. 18 at www.bbc.co.uk/news/world-south-asia-10896849

Bempah, S. and Øyhus, A. (2017). The role of social perception in disaster risk reduction: Beliefs, perception, and attitudes regarding flood disasters in communities along the Volta River, Ghana, *International Journal of Disaster Risk Reduction*, 22, 104–108.

Boamah, S., Armah, F., Kuuire, V., Ajibade, I., Luginaah, I. and McBean, G. (2015). Does experience of floods stimulate the adoption of coping strategies? Evidence from cross sectional surveys in Nigeria and Tanzania, *Environments*, 2, 565–585.

Briscoe, J. (2009). Water security: Why it matters and what to do about it, *Innovations*, 4(3), 3–28.

Cumiskey, L., Werner, M., Meijer, K., Fakhruddin, S. and Hassan, A. (2015). Improving the social performance of flash flood early warnings using mobile services, *International Journal of Disaster Resilience in the Built Environment*, 6(1), 57–72.

Gerlak, A., Lautze, J. and Giordano, M. (2011). Water resources data and information exchange in transboundary water treaties, *International Environmental Agreements: Politics, Law and Economics*, 11(2), 179–199.

Hallegatte, S., Bangalore, M., Bonzanigo, L., Fay, M., Kane, T., Narloch, U., Rozenberg, J., Treguer, D., and Vogt-Schilb, A. (2016). *Shock Waves: Managing the Impacts of Climate Change on Poverty*, World Bank, Washington, DC.

Hossain, F. (2007). Satellites as the panacea to transboundary limitations for longer term flood forecasting? *Water International*, 32(3), 376–379.

ICIMOD (2018). Establishment of a regional flood information system in the Hindu Kush Himalayan region (HKH-HYCOS). Accessed at www.icimod.org/?q=264.

Keoduangsine, S. and Goodwin, R. (2012). An appropriate flood warning system in the context of developing countries, *International Journal of Innovation, Management and Technology*, 3(3), 213–216.

Lamond, J. and Proverbs, D. (2009). Resilience to flooding: Lessons from international comparison, *Urban Design and Planning*, 162(2), 63–70.

MacClune, K., Venkateswaran, K., Dixit, K., Yadav, S. and Maharjan, R. (2015). Urgent case for recovery: What we can learn from the August 2014 Karnali River floods in Nepal, ISET International, ISET-Nepal and Practical Action, Nepal, and Zurich. Accessed at https://reliefweb.int/sites/reliefweb.int/files/resources/risk-nexus-karnali-river-floods-nepal-july-2015.pdf.

Mahmud, I., Akter, J. and Rawshon, S. (2012). SMS-based disaster alert system in developing countries: A usability analysis, *International Journal of Multidisciplinary Management Studies*, 2(4), 1–15.

Mileti, D. (1995) Factors related to flood warning response, U.S.-Italy Research Workshop on the Hydrometeorology, Impacts, and Management of Extreme Floods, November, Perugia, Italy.

Molden, D., Harma, E., Shrestha, A., Chettri, N., Pradhan, S. and Kotru, R. (2017). Focus Issue: Implications of out and in migration for sustainable development in mountains, *Mountain Research and Development*, 37(4), 502–508.

Niang, M. (2016). Senegal floods expose need for community warning, preparation, Reuters, London. Accessed at www.reuters.com/article/us-senegal-disaster-floods-warning/senegal-floods-expose-need-for-community-warning-preparation-idUSKCN10Q0M8.

Ngwese, N., Saito, O., Sato, A., Boafo, Y, and Jasaw, G. (2018). Traditional and local knowledge practices for disaster risk reduction in Northern Ghana, *Sustainability*, 10, 1–17.

Parker, D. J. (2003). Designing flood forecasting, warning and response systems from a societal perspective, International Conference on Alpine Meteorology and Meso-Alpine Programme, Brig, Switzerland.

Parker, D.J. and Handmer, J. (1998). The role of unofficial flood warning systems, *Journal of Contingencies and Crises Management*, 4(1), 45–60.

Parker, D. J. (2017). Flood warning systems and their performance, *Oxford Research Encyclopedia of Natural Hazard Science*, Oxford University Press, Oxford.

Paul, B., Rashid, H., Islam, M., and Hunt, L. (2011). Cyclone evacuation in Bangladesh: Tropical cyclones Gorky (1991) vs. Sidr (2007), *Environmental Hazards*, 9(1), 89–101.

Paul, S. and Routray, J. (2010). Flood proneness and coping strategies: the experiences of two villages in Bangladesh, *Disasters*, 34(2), 489–508.

Perez, R., Espinueva, S. and Hernando, H. (2007). Community-based flood early warning systems. Briefing Paper: Workshop on the science and practice of flood disaster management in urbanizing Monsoon Asia, Philippine Atmospheric, Geophysical and Astronomical Services Administration (PAGASA), Chiang Mai, Thailand.

Plate, E. (2007). Early warning and flood forecasting for large rivers with the lower Mekong as example, *Journal of Hydro-environment Research*, 1, 80–94.

Plengsaeng, B., When, U. and van der Zaag, P. (2014). Data-sharing bottlenecks in trans-boundary integrated water resources management: A case study of the Mekong River Commission's procedures for data sharing in the Thai context, *Water International*, 39(7), 933–951.

PAMC (Practical Action and Mercy Corps) (2012). Community-based early warning systems in South and South East Asia, PAMC, Rugby, UK.

Rahman, M., Goel, N. and Arya, D. (2012). Study of early flood warning dissemination system in Bangladesh, *Journal of Flood Risk Management*, 6, 290–301.

Rogers, D. and Tsirkunov, V. (2013). Weather and climate resilience: Effective preparedness through national meteorological and hydrological services, Directions in Development, World Bank, Washington, DC.

Samaddar, S., Misra, B. and Tatano, H. (2012). Flood risk awareness and preparedness: The role of trust in information sources, 2012 IEEE International Conference on Systems, Man, and Cybernetics, Seoul, Korea.

Schipper, E. (2010). Religion as an integral part of determining and reducing climate change and disaster risk: An agenda for research. In: Voss, M. (ed.), *Der Klimawandel. VS Verlag für Sozialwissenschaften*, Springer Nature, Switzerland, 377–393.

Sha, M., Douven, W., Werner, M. and Leentvaar, J. (2012). Flood warning responses of farmer households: A case study in Uria Union in the Brahmaputra flood plain, Bangladesh, *Journal of Flood Risk Management*, 5, 258–269.

Shrestha, M., Goodrich, C., Udas, P., Rai, D., Gurung, M. and Khadgi, V. (2016). Flood early warning systems in Bhutan: A gendered perspective, ICIMOD Working Paper 2016/13, ICIMOD, Kathmandu.

Theimig, V., de Roo, A., and Gadain, H. (2011). Current status on flood forecasting and early warning in Africa, *International Journal of River Basin Management*, 9(1), 63–78.

Thu, N. and When, U. (2016). Data sharing in international transboundary contexts: The Vietnamese perspective on data sharing in the Lower Mekong Basin, *Journal of Hydrology*, 536, 351–364.

van Voorst, R. (2014). The right to aid: Perceptions and practices of justice in a flood-hazard context in Jakarta, Indonesia, *Asia Pacific Journal of Anthropology*, 15(4), 339–356.

Webster, P. and Jian, J. (2011) Environmental prediction, risk assessment and extreme events: Adaptation strategies for the developing world, *Philosophical Transactions of the Royal Society*, 369, 1–30.

World Bank (2010). Natural hazards, unnatural disasters: The economics of effective prevention, World Bank, Washington, DC.

World Bank (2016). Modernizing meteorological services to build climate resilience across Africa. Accessed at www.worldbank.org/en/news/feature/2016/11/10/modernizing-meteorological-services-to-build-climate-resilience-across-africa

WMO (World Meteorological Organisation) (2008). Capacity assessment of national meteorological and hydrological services in support of disaster risk reduction, WMO, Geneva.

WMO (World Meteorological Organisation) (2017). Niger making progress towards a flood early warning system, WMO, Geneva. Accessed at https://public.wmo.int/en/media/news/niger-making-progress-towards-flood-early-warning-system

WMO (World Meteorological Organisation) (2018). Multi-hazard Early Warning Systems: A checklist. Outcome of the first Multi-hazard Early Warning Conference, WMO, Geneva.

WRI (World Resources Institute) (2015). World's 15 countries with the most people exposed to river floods, WRI, Washington, DC. Accessed at www.wri.org/resources/maps/aqueduct-global-flood-analyzer

Zia, A. and Wagner, C. (2015). Mainstreaming early warning systems in development and planning processes: Multilevel implementation of Sendai framework in Indus and Sahel, *International Journal of Disaster Risk Science*, 6, 189–199.

11

FLOOD WARNING AND RECOVERY IN ZIMBABWE

Some salutary lessons

Abigail Tevera

Introduction

Flood warning systems have 'chain-like' characteristics (Figure 11.1; see also chapter 10) and are therefore liable to frailty because the whole chain is only as strong as the weakest link (Parker and Priest, 2012). This chapter explores the effectiveness of a flood warning in Zimbabwe, using the case of the recent Tokwe-Mukorsi flood, building on results from interviews conducted with key stakeholders in Zimbabwe, analysed to evaluate the effectiveness of the flood warning system involved.

Aims and practices

A flood warning system should allow national and local authorities to make decisions on protecting the lives of the public, infrastructure, properties, the environment and agriculture, as well as planning for aid that may be needed before and after the flood event (Alfieri et al., 2012; Pappenberger et al., 2015). It involves three main elements: flood identification and mapping; flood risk information provision; and emergency response preparedness (Parker and Priest, 2012).

Timeliness, accuracy and reliability are prerequisites for an adequate flood warning and to ensure an appropriate response. If people situated in flood-affected areas do not get the right information about the flood risks on time, in a form that is clear and easy to understand, in order to facilitate them to evacuate and move from their properties before the flood arrives, there will be a higher risk of damage to key infrastructure, people's lives and properties.

However, this is not always achievable because the degree of success of a flood warning system (FWS) is dependent on several technical and human factors which may have deficiencies that are propagated through the whole interwoven process.

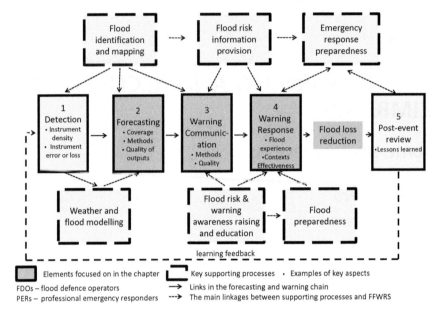

FIGURE 11.1 The flood forecasting, warning and response chain and key supporting processes

Source: reproduced with permission, Parker and Priest (2012).

'Most early warning systems fail to reach the end users due to bureaucratic protocols, and as a result don't serve the purpose of early response and evacuation' (Shrestha et al., 2014). While technological limitations and indeed failure can occur, despite many recent scientific advances, human factors and related social issues appear to be the most likely potential deficiencies that define the ultimate degree of success of flood forecasting, warning and response arrangements.

Barriers to effective flood risk warning systems include, first, the difficulties in translating scientific flood forecasting information into a simple format that can be transmitted easily to the relevant professionals and to the people at risk of flooding (Faulkner et al., 2007). Second, flood risk communication systems are still facing technological challenges, especially in developing countries due to their poor economic status and hence inadequate investment (see also chapter 10). In a situation close to Zimbabwe, for example, the radar equipment that is currently used in South Africa is old and 'bad telephone communication between measuring stations and the SA Weather Service still featured during floods'; these therefore can be the ultimate reason for no effective response to flood forecasts (Du Plessis, 2002, 135). Recent findings (Lumbroso et al., 2016) also highlighted lack of capacity and funding as one of the six main barriers preventing the development of effective early warning systems and flood risk assessments in Africa.

Third, even if scientists carry out excellent flood detection and flood forecasting and the relevant people on the FWS chain receive accurate forecast information which is on time, clear and can be sensibly translated and disseminated as flood warnings, the whole system will be liable to break down and fail completely if the public does not respond accordingly. The Hyogo Framework for Action 2005–2015 from the United Nations Office for Disaster Reduction (UNISDR, 2005) emphasises the need for the development of early warning systems that are people-centred, with the demographic characteristics of the target public taken as one of the four key considerations in managing disasters. More broadly, establishing good communication networks is one effective way found to increase people's response to flood warnings within a local community at risk of flooding (Parker et al., 2009). The flood plain occupants can learn about flood risks from their local networks to help them prepare for emergencies (Parker and Handmer, 1998).

The Tokwe-Mukorsi flood disaster

Background

In February 2014, Zimbabwe's huge Tokwe river basin flooded, following heavy rains, affecting the infrastructure and properties of members of 12 villages upstream of the Tokwe-Mukorsi Dam in Masvingo Province (UNICEF, 2014). Well over 20,000 people had to be evacuated (Defend Truth, 2014) and, in addition, people appear to have died as a result of the flood disaster (Moyo, J., 2014; Moyo, M., 2014; Samukange, 2014). It is against this backdrop of this type of losses and community dislocation that we question whether the flood warning system in Zimbabwe, in this case, was effective and, if not, why not.

We rely here on accounts given of the flood disaster using first-hand information from key informants, literature written by scholars and media articles on this incident. Researchers choose key informants to study organisations because they are knowledgeable about the issues being researched and are able and willing to communicate about them (Kumar et al., 1993). 'The principal advantages of the key informant technique relate to the quality of data that can be obtained in a relatively short period of time' (Marshall, 1996, 2).

However, in common with other research techniques, this methodology also has its weaknesses, which include but are not limited to the fact that there is a possibility that those in government or in quasi-governmental agencies (such as we contacted here (Table 11.1)) may not disclose information that they feel could be politically sensitive. Accounts also may be somewhat conflicting and irreconcilable. We need to bear in mind these caveats when interpreting the results below.

The flood event

The Tokwe-Mukorsi Dam is a concrete-faced rock-filled structure on the Tokwe River, just downstream of its confluence with the Mukorsi River, about 72 km south of Masvingo in Masvingo Province, Zimbabwe. At the time of the flood the dam was still under construction.

TABLE 11.1 Key informant interviewees

Warning system elements	Roles/functions	Respondents
(1) Detection (flood identification and mapping)	• Hydrologists/Flood modellers	• Representative of the Zimbabwe National Water Authority (ZINWA)
(2) Forecasting (flood risk information provision)	• Meteorologists/Weather and flood forecasters	• Representative of the Zimbabwe Meteorological Services Department (MSD)
(3) Warning communication and response (including emergency response preparedness)	• Emergency Service Manager • Local authorities • Non-Governmental Organisation (NGO)	• The Provincial Administrator • The District Administrator • Representative of UNICEF

In late 2013, the government of Zimbabwe, after receiving the first flood alert from the Zimbabwe National Water Authority (ZINWA), planned to relocate the people that were at risk from the forecasted flooding. The government had previously put in place a plan and structure for relocating the communities that were situated in the area to be flooded when the dam was completed, a process due to be carried out over time in three phases. The first group of communities, consisting of 611 families, were indeed safely relocated to Nuanetsi ranch at Masangula and Chisase.

However, the dam area received double the normal rainfall in early 2014 (850 mm) (Gumindoga et al., 2014) easily surpassing the anticipated annual total here of 660 mm. Water levels rose markedly in the basin, culminating in flooding of the area and causing a partial failure on the downstream face of the dam. The valley behind the dam had been expected to fill gradually over four years, with that plan to relocate people in phases, but it did so over a matter of weeks (Gumindoga et al., 2014) resulting is sudden inundation (UNOSAT, 2014). The 12 villages within the basin were caught unawares.

As a result, a total of 2,514 households living upstream of the dam were directly affected by the flooding (UNICEF, 2014). The government's previous phased plans were disrupted because of the unanticipated level of heavy rain and flooding which was not communicated to the 6,393 families that were living in the area around the dam by the relevant authorities responsible for the detection and forecasting of such events.

In this case, the government assessed that these 6,393 families would need to be relocated as an emergency, and it designated three relocation sites: (1) Chisase in Masvingo district, 50 kilometres from the dam (126 families; one school, one clinic, 42 boreholes); (2) Masangula in Mwenezi district, 150 kilometres from the dam (485 families; three schools, two clinics; 42 boreholes); and lastly Chingwizi in Mwenezi district, 150 kilometres from the dam (5782 families; five schools; two clinics; 63 boreholes) (Mavhura et al., 2017; Human Rights Watch, 2015). On 9 February 2014, the president of the Zimbabwe declared this incident a national disaster and mobilised more resources to assist the affected families.

Discussion of results and analysis

Flood risk identification and mapping

In Zimbabwe, two departments, the Meteorological Services Department (MSD) and ZINWA, are responsible for monitoring and advising on weather conditions, with ZINWA being responsible from a hydrological perspective and MSD from a weather and climate information or services perspective.

In the case of the Tokwe-Mukorsi flood disaster, ZINWA used the following techniques to detect the flood, as indicated by the representative of that authority who we interviewed:

- Flood modelling using in-house software;
- Stream gauge monitoring;
- Upstream dam monitoring;
- GIS and remote sensing.

According to our ZINWA respondent, although the data analysis, risk prediction and warning generation was based on accepted scientific and technical methodologies, she felt more could have been done in terms of the capacity building of the team to ensure effective early warning systems. Capacity building is required in the form of training more staff and replacing outdated arrangements with modern equipment. This is because there were not enough adequately trained staff to use the tools for prediction and forecasting for this flood.

The flood prediction and forecasting information that was produced was shared with the Civil Protection Unit (CPU) members and the catchment office. It was also shared in the form of hydrological bulletins and through presentations at meetings and by email with stakeholders including the Zimbabwe air force, MSD, ZINWA, the Zimbabwe Republic Police (ZRP), research institutions and NGOs. Although relevant techniques for flood detection and forecasting were employed, some of the technology had major constraints in not being reliable and therefore delivering erroneous outputs.

Warnings were dependent on the MSD forecasts within Zimbabwe. The MSD role, according to the MSD representative we interviewed, is to provide accurate and timely weather and climate information or services for all events from now-casting to sub-seasonal and seasonal weather and longer-term climate predictions. The department uses remote sensing technology (satellite imagery), numerical weather prediction models and synoptic data for monitoring the evolution of dynamic or mesoscale systems responsible for the genesis of thunderstorms, and the Southern Africa Flash Flood Guidance System (SARFFG) to predict and generate flood warnings in Zimbabwe.

From the information provided by the MSD interviewee, it appears that the department is well equipped to detect and forecast weather events, but the question remains as to whether all relevant stakeholder groups could access the information.

According to our interviewee, the flood warning information did not reach all the relevant stakeholders and, most importantly, the communities at risk due to communication gaps in the means through which this information would be disseminated quickly. The prediction and forecasting information was shared through press releases (warnings and alerts) as well as press briefings, but, for example, not all communities had access to radios. Neither were there any associated training programmes or mock drills in place to make all stakeholders understand the prediction and forecasting information they generated and shared with them (the provincial administrator; district administrator; NGOs; community members at risk). If the right techniques were used and the information was translated into formats that the stakeholders could understand, and if the right channels were followed in transmitting the flood risk warning, the question remains, therefore, why did the flood so seriously affect the 2,514 households living upstream of Tokwe-Mukorsi Dam.

It seems inevitable that there must have been a deficiency somewhere in the flood warning chain. In this respect it would appear that there were issues related to the legal responsibility for the warnings and their transmission/dissemination. Part VIII, Section 27 of the Civil Protection Act states that the head of state is the only person who can declare a state of disaster after having received recommendations from the responsible minister. In this case it was reported that water levels rose dramatically on 27 January 2014 (Gumindoga et al., 2014), but the declaration of the disaster that would necessitate a proper flood response was only made on 9 February 2014.

Therefore, this lack of institutional/legal capacity to communicate on such serious flooding risk to the relevant people at this stage becomes one of the points of weakness in the Zimbabwean flood warning chain. Another conclusion that can be drawn is that the early warning failed at this stage because the flood detection and forecasting equipment was also defective and therefore not reliable. This point is supported by findings from research by Zirecho (2016) which suggested that the Zimbabwean government was let down by the Meteorological Services Department (MSD): no early warning signal was provided prior to the floods as the equipment there was outdated, and they could only provide a warning with a short lead time between the warning and the danger.

Flood risk information provision

The provincial administrator has the legal responsibility for flood warnings and their dissemination to the public. However, the nature of the legal system itself poses a big challenge to timely risk communication to the public because the provincial administrator cannot take any action until the president has officially made his declaration.

In the case of the Tokwe-Mukorsi flood disaster, the response we obtained from the provincial administrator on this issue was that they were not informed of any possible occurrence of floods and learnt of the disaster through phone calls from victims whose homes were submerged in water. After finally receiving the declaration of disaster from the president on 9 February, the provincial adminis- trator then transmitted it to the floodplain dwellers through the district adminis- trator, village leaders, the police sub-aqua team and the media, although by that time it was already too late.

This dysfunction of the flood warning system was also highlighted by Zirecho (2016), in that all his academic and civil society interviewees pointed out that it was because of the dysfunction of the Civil Protection Unit (CPU) that the floods had caught many people unprepared in this area. Also, 'It is apparent that the Tokwe-Mukorsi flood victims' miserable experiences were a result of the govern- ment's lack of adequate planning and lack of foresight ...' (Hove, 2016, 11). The highly centralised decision-making process in flood management is one of the weaknesses of the flood warning system in Zimbabwe (Gwimbi, 2007). The weak link between the CPU and the provincial administrator did not allow the evacua- tion of the flood victims in time or the government to respond accordingly by compensating the flood victims and resettling them in new areas, hence the big losses.

Emergency response and recovery

Our information from the district administrator and the humanitarian organisation UNICEF (Table 11.1) showed that poor flood response was in this case a result of delays in communicating flood risk and lack of access to information on that risk by the people who were liable to be flooded. The district administrator confirmed that, at the local level, they had not received any flood warning information from ZINWA or the provincial administrator before the Tokwe-Mukorsi disaster occurred.

They were caught unawares and were therefore not able to carry out any flood risk awareness campaigns for those at risk or to put in place any associated training programmes or mock drills before the flood hit the area. The UNICEF repre- sentative also noted the same constraints, leading to a great loss of property, live- stock and even human life in the flood disaster (see also Samukange (2014)). They also confirmed that the government, represented by the CPU, was reluctant to provide the relevant warning information until the event was declared a national disaster by the president.

But he also suggested that in his view there was no evidence that people would have used the flood warning information even if it had been provided: 'People continued to stay in their homes until they were in the floods', he wrote. He added that UNICEF only knew about the flooding after it had occurred and communities had already been displaced. A flood victim interviewed by a repre- sentative of the International Federation of Red Cross and Red Crescent Societies

(Hansika Bhagani, 2014) said: 'We had been told about the impending floods but they came too fast so we could not prepare. We lost our ploughing tools for farming, our beds, because those could not fit into the helicopter. I am very worried about our livestock'.

In terms of the last link of the chain discussed above (the post-event review), Kadzatsa (2014, 5) provides a strategic overview of some of the lessons learned by the government of Zimbabwe, the disaster risk management stakeholders and their partners from the rescue and relief activities in the immediate aftermath of the Tokwe-Mukorsi flood disaster. He concluded from a widely represented workshop with more than 80 participants, held in November 2014, that those search and rescue operations were effective but the relocation to permanent residence and early recovery measures were areas that experienced serious challenges. The rapid nature of the flood disaster – and a clearly deficient warning and information dissemination system – prevented the orderly and planned resettlement of the people displaced by the dam project and affected by the flood. A key workshop conclusion – among many – was therefore that enhanced flood monitoring including real-time monitoring and dissemination facilities for early warning information and flood modelling and mapping should be prioritised (Kadzatsa, 2014, 10–11).

Conclusions

Social and political problems define the success of a flood warning system and therefore cooperation is required between the government, relief agencies and communities to create, maintain and use that system for it to function properly (Basha and Rus, 2007).

The case study presented here revealed many narratives regarding the ineffectiveness of a Zimbabwean flood warning system. Previous research by Gwimbi (2007) had identified similar weaknesses including a fragmented approach to flood management and a lack of local community involvement, especially in the decision-making process.

Because the three flood warning system elements described above are connected to each other in the form of a chain, each element of that chain needs to perform efficiently to yield an effective flood warning system. This did not happen in the case of the Tokwe-Mukorsi flood disaster. The communication and dissemination components did not receive adequate attention and this resulted in a huge gap between the information produced by ZINWA/MSD and the information that was received and acted upon by the other stakeholders: the district administrator, the humanitarian organisations, and the affected communities.

Learning from this example (cf. Kundzewicz, 2013), we can conclude that carefully demarcated decentralisation of decision-making is a key component in helping to avoid delays in communicating the danger to the people at risk of flooding and allow for a timely response by everyone involved (the flood plain dwellers, and also the government, in terms of compensation and relocation of the victims). The political will of governments to invest in post-event reviews to facilitate that learning is indispensable so that lessons from individual flood events can be used to improve flood risk management in future situations, and this is perhaps especially the case in developing countries such as Zimbabwe.

References

Alfieri, L., Salamon, P., Pappenberger, F., Wetterhall, F. and Thielen, J. (2012). Operational early warning systems for water-related hazards in Europe. *Environmental Science & Policy*, 21, 35–49.

Basha, E. and Rus, D. (2007). Design of early warning flood detection systems for developing countries. In *2007 International Conference on Information and Communication Technologies and Development*. IEEE, 1–10.

Defend Truth (2014). In photos: The aftermath of Zimbabwe's Tokwe-Mukorsi dam disaster. Available at: www.dailymaverick.co.za/article/2014-04-14-in-photos-the-afterma th-of-zimbabwes-tokwe-mukorsi-dam-disaster/#.VKVoZivF-So. [Accessed 2 August 2018]

Du Plessis, L. (2002). A review of effective flood forecasting, warning and response system for application in South Africa. *Water SA*, 28, 129–138.

Faulkner, H., Parker, D. J., Green, C. H. and Beven, K. (2007). Developing a translational discourse to communicate uncertainty in flood risk between science and the practitioner. *Ambio*, 36, 692–704.

Gumindoga, W., Chikodzi, D., Rwasoka, D., Mutowo, G., Togarepi, S. and Dube, T. (2014). The spatio-temporal variation of the 2014 Tokwe-Mukorsi floods: a GIS and remote sensing-based approach. *Journal of Science Engineering and Technology, Zimbabwe Institution of Engineers*, 1(2), 1–10.

Gwimbi, P. (2007). The effectiveness of early warning systems for the reduction of flood disasters: Some experiences from cyclone induced floods in Zimbabwe. *Journal of Sustainable Development in Africa*, 9, 152–169.

Hansika Bhagani (2014). Heavily pregnant, scared and displaced in flooded Zimbabwe. International Federation of Red Cross and Red Crescent Societies (IFRC). Available at: www.ifrc.org/en/news-and-media/news-stories/africa/zimbabwe/heavily-pregnant-sca red-and-displaced-in-flooded-zimbabwe-65227/.

Hove, M. (2016). When flood victims became state victims: Tokwe-Mukorsi, Zimbabwe. *Democracy and Security*, 12, 135–161.

Human Rights Watch (2015). Homeless, landless, and destitute: The plight of Zimbabwe's Tokwe-Mukorsi flood victims. Available at: www.hrw.org/report/2015/02/03/homeles s-landless-and-destitute/plight-zimbabwes-tokwe-mukorsi-flood-victims.

Kadzatsa, M. (2014). Tokwe-Mukorsi flood disaster: Lessons learned workshop. Report. Department of Civil Protection (and others,) Masvingo, Zimbabwe. Available at: www. drmzim.org/wp-content/uploads/2015/06/Tokwe-Mukorsi-Lessons.pdf.

Kumar, N., Stern, L. W. and Anderson, J. C. (1993). Conducting interorganizational research using key informants. *Academy of Management Journal*, 36, 1633–1651.

Kundzewicz, Z. (2013). Floods: Lessons about early warning systems. In: EEA (European Environment Agency) (ed.), *Late lessons from early warnings: Science, precaution, innovation*, 347–368. Luxembourg: Publications Office of the European Union.

Lumbroso, D., Brown, E. and Ranger, N. (2016). Stakeholders' perceptions of the overall effectiveness of early warning systems and risk assessments for weather-related hazards in Africa, the Caribbean and South Asia. *Natural Hazards*, 84, 2121–2144.

Marshall, M. N. (1996). The key informant technique. *Family Practice*, 13, 92–97.

Mavhura, E., Collins, A. and Bongo, P. P. (2017). Flood vulnerability and relocation readiness in Zimbabwe. *Disaster Prevention and Management*, 26, 41–54.

Moyo, J. (2014). Zimbabwe: Villagers cornered by flood disaster [Online]. *Allafrica.* Available at http://allafrica.com/stories/201402241728.html. [Accessed 5 January 2017]

Moyo, M. (2014). Zimbabwe: Flash flood update #1 (as of 7 February 2014). United Nations Office for the Coordination of Humanitarian Affairs, New York.

Pappenberger, F., Cloke, H. L., Parker, D. J., Wetterhall, F., Richardson, D. S. and Thielen, J. (2015). The monetary benefit of early flood warnings in Europe. *Environmental Science and Policy*, 51, 278–291.

Parker, D. J. and Handmer, J. W. (1998). The role of unofficial flood warning systems. *Journal of Contingencies and Crisis Management*, 6, 45–60.

Parker, D. J. and Priest, S. J. (2012). The fallibility of flood warning chains: can Europe's flood warnings be effective? *Water Resources Management*, 26, 2927–2950.

Parker, D. J., Priest, S. J. and Tapsell, S. M. (2009). Understanding and enhancing the public's behavioural response to flood warning information. *Meteorological Applications*, 16, 103–114.

Samukange, T. (2014). Tokwe-Mukorsi –is Zimbabwe sleepwalking? *Newsday*, 17 February. Accessed at: www.newsday.co.zw/.

Shrestha, M., Kaphle, Ś., Gurung, M. B., Nibanupudi, H. K., Khadgi, V. R. and Rajkar-nikar, G. (2014). Flood early warning systems in Nepal: A gendered perspective. International Centre for Integrated Mountain Development, Kathmandu, Nepal. Available at: http://lib.icimod.org/record/29959/files/Flood_EWS.pdf.

UNICEF (2014). UNICEF Zimbabwe CO situation report # 2. UNICEF, Harare.

UNISDR (2005). Hyogo framework for action 2005–2015 [Online]. Available at: www.unisdr.org/2005/wcdr/intergover/official-doc/L-docs/Hyogo-framework-for-actio n-english.pdf. [Accessed 3 January 2017]

UNOSAT (2014). Flood water over Tokwe-Mukorsi Dam, Masvingo Province, Zimbabwe. UNOSAT/UNITAR, Meyrin, Switzerland.

Zirecho, S. (2016). Effectiveness of international response mechanisms to national disasters: the case of Tokwe-Mukorsi 2014 floods in Zimbabwe. Unpublished doctoral dissertation, Bindura University of Science Education. Available through: http://humanities2139.rssing.com/chan-61507118/all_p1.html.

12

ADAPTING TO FLOODS IN SOCIAL HOUSING IN THE UK

A social justice issue

Diana King and Edmund C. Penning-Rowsell

Introduction

The British government has a history of intervening to protect those at risk from flooding, most recently through the policy of flood risk management (FRM) (Penning-Rowsell and Johnson, 2015). This shift towards FRM from flood defence reflects the changing attitudes regarding flooding towards its risks being managed in a sustainable and equitable manner (Johnson et al., 2007) rather than attempting to eliminate them.

The uptake of the concept of sustainability in FRM has been fairly uncontentious, with a greater focus on improving non-structural FRM policies (e.g. insurance; spatial planning; flood warnings). This has accompanied a shift of some responsibilities from the state to the individual to address flood risk (Harries, 2012), so that communities are encouraged to understand how to 'live with floods' (Johnson et al., 2008, 1). However, the uptake of equity in FRM policies has been caught in a debate on social justice (O'Neil and O'Neil, 2012; Sayers et al., 2017).

For centuries, philosophers have debated the definition of justice, with each philosophy producing different principles as to what a just, 'fair' or equitable outcome should look like (Johnson et al., 2005). These range from providing equal support to everybody, regardless of risk, to prioritising the most vulnerable. In using these social justice principles, it is possible to analyse how equitable the FRM policies might be. However, it is also important to recognise the differing capacities of residents to engage with these policies. In this context we evaluate here how socially just some key FRM policies are for social housing residents, who are some of the most disadvantaged in the UK (Shelter, 2014).

Social justice and flood risk management

The biggest issue in transferring the concept of equity into water policy concerns the multiple definitions and interpretations of that concept. We use three of the most influential philosophies to demonstrate the differing outcomes of justice and the implications of this for FRM policies (Johnson et al., 2005).

First, the *egalitarian* philosophy favours equality, from ensuring that resources are distributed equally to providing equal access of opportunity. Egalitarians focus on individual capabilities (e.g. individuals are able to participate in politics; access to health services) to assess whether the system is producing fair opportunities for everyone (Sen, 1992) (i.e. is procedurally just). As noted by Johnson et al. (2007), an egalitarian FRM policy would 'ensure that all those at risk of flooding have an equal opportunity of having their flood risk managed by the state'.

In contrast, second, *Rawlsian* justice focuses on the institutions that provide services and resources, stating that they have the duty only to allow inequalities if the policy provides the greatest benefit to the least advantaged (Rawls, 2001). For FRM, a Rawlsian philosophy would prioritise distributional justice (i.e. a fair distribution of resources, and good and bad outcomes), ensuring that resources are directed to the most vulnerable to flood risk.

Third, a *utilitarianism* philosophy also focuses on the principle of distributional justice. Yet, utilitarian beliefs aim to redistribute societal resources to maximise potential societal 'happinesses' (Mill, 1863), often by using a benefit:cost approach. Utilitarians would promote FRM policies that maximise total utility and are thus beneficial to the entire country. A critique of utilitarian beliefs is that maximising utility fails to consider the uneven distribution of needs and thus differentiated vulnerabilities. However, utilitarian thinking would state that as FRM is funded by all taxpayers and not just those at risk of flooding, the investments should benefit all taxpayers and be as economically efficient as possible (Johnson et al., 2008).

If FRM policies are to be considered equitable and fair, they will likely exemplify one of the above philosophies of justice, either by aiming to provide support for everyone (egalitarian), or to help the most disadvantaged (Rawlsian), or to maximise returns for the nation as a whole (utilitarian). These three principles provide useful criteria by which to assess in this chapter the extent to which particular non-structural FRM policies are socially just.

Vulnerability and capability with regard to social housing

We see vulnerability to environmental risk as a function of three factors: the *exposure* and *sensitivity* to the biophysical risk, and the *adaptive capacity* to cope with, respond to and prepare for that risk (Smit and Wandel, 2006).

This first implies that vulnerability will be spatially clustered throughout the UK, as a result of the environmental characteristics that produce patterns of exposure (e.g. coastal areas; floodplains). Second, even if exposure to flood risk is equal within a community, the impacts felt at the household level will be dependent on social factors (e.g. income; education) and personal attributes (e.g. age; health) (Lindley et al., 2011). We therefore need to consider social housing, with its correlation with household income, before characterising the vulnerability (and capacity) of its residents to cope with and adapt to flood risk.

Social housing first emerged in the UK during the nineteenth century to meet the housing deficit that accompanied the urban population boom with the industrial revolution. The 1919 Housing Act mandated local municipalities, with the support of subsides from central government, to produce safe houses for working class families (Hollow, 2011). This social housing was initially rented. However, government subsidies have since encouraged tenants to buy a part of (i.e. Social HomeBuy legislation) or all of (i.e. Right to Buy) their dwellings (UK Government, 2014; Glynn, 2009). In the 1990s, the Labour government successfully encouraged the shift of much of the management of housing from the state to not-for-profit Housing Associations (Blair, 1998), while introducing the Decent Homes standard in 2000 for that social housing (Whitehead and Scanlon, 2007).

Over the past century, the differing political objectives for social housing have resulted in a heterogeneous stock. The property size (e.g. of a room), type (e.g. apartment or semi-detached) and quality (e.g. of building materials) are all factors that depend on the era in which each unit was built (Whitehead and Scanlon, 2007); this will produce differentiated vulnerabilities to flooding. For example, the impacts of flooding are usually concentrated on the ground floor, affecting fewer people in a tower block than an estate of semi-detached properties (Pringle et al., 2013). Furthermore, the varied ownership structures have produced unclear directives on who is responsible for maintaining a particular dwelling, reducing the capacity of residents to flood-proof their houses or repair them after the flooding.

Individuals and families in-need (e.g. those who are homeless) are given *reasonable preference* during the allocation of social housing (Shelter, 2014). Half of social housing tenants (vs. 14% in other housing types) are categorised as low-income households as they earn 60 per cent or less than the British median (The Poverty Site, 2014). Social housing residents are disproportionally both old and young and tend to attain low levels of education (Taylor, 1998).

According to the vulnerability approach, social housing residents are likely to be more susceptible to flooding through a limited capacity (e.g. financial assets; mobility) to cope with disturbances. However, while social deprivation and social housing may overlap (Taylor, 1998) to produce spatially distributed vulnerability to flooding, not all tenants in social housing have similar characteristics. For example, in 2012 there were between 1,000 and 6,000 households in social housing in England where the total income exceeded £100,000 per annum, and 12,000 to 34,000 earning at least £60,000 (DCLG, 2012). Therefore, given that income buys resilience, we need to note that not all tenants in social housing are equally vulnerable to any of the risks that they face.

Social housing, flood risk management, and social justice

One in six British properties is currently judged by the government to be exposed to some flood risk (Environment Agency, 2009) although the exact number is unclear (Penning-Rowsell, 2015). Structural defences cannot be used to protect all these properties, primarily for economic reasons (Johnson et al., 2005). Non-structural

FRM policies build local resilience to flooding at a somewhat lower cost, not least to the state. Yet, such policies often require household level action for them to be effective. Using our three criteria of justice alongside the characterisation of vulnerability for social housing residents (p. xxx), we now examine how socially just are three of these interventions before discussing two case studies relevant to the same themes.

Flood awareness and warnings

Public awareness campaigns happen well before a flood might occur, to promote appropriate behaviour when flooding comes. Flood warnings are given out closer to the flood event, to help communities prepare for the imminent flood.

In England, the Environment Agency is responsible for achieving both of these aims. They recently restated their commitment to providing near-universal and equal coverage for all UK residents (Environment Agency, 2014), in line with the distributional justice heralded by egalitarians. The Environment Agency also aims to 'target resources to reduce the risk of flooding to communities with the highest flood risk' (Swindon Borough Council, 2014, 14), demonstrating an attention to differentiated vulnerability and the promotion of Rawlsian fairness.

Yet there is evidence that the effectiveness of these policies for social housing tenants depends on their level of income, education and language ability (Pringle et al., 2013). Research also indicates that low-income households are less likely to know how to cope with flooding (Fielding and Burningham, 2005). As such, while this policy may pass two of our equity criteria (egalitarian; Rawlsian), it still may in practice fail to engage part of their audience and reduce their flood risk.

Home adaptations

Home adaptations are an effective way of increasing resistance and resilience to flooding, whether with coping mechanisms (e.g. moving belongings), immediate actions (e.g. using sandbags), or future anticipatory adaptations (e.g. raising exterior door thresholds). While some of these options are cost-free, others require the purchase of products, and may be thus less available to low-income households (Porter et al., 2014).

The UK's Decent Homes policy provides a standard of conditions for social housing, which is Rawlsian in nature as it directs resources to some of the most deprived households in the UK. Yet, these standards do not necessarily accommodate current or future flood risk (Sustainable Homes, 2013). Furthermore, as contractors are used to undertake this work, it is unclear whether tenants have the authority to make structural changes to their dwellings, should they wish to carry out additional flood-proofing. This lack of autonomy has implications for a tenant's ability to undertake household adaptations to reduce flood risk, first by creating apathy for action (i.e. through tenants not knowing, and deferring action to the manager (Grothmann and Reusswig, 2006), and second through refusing tenants the right to act (i.e. being obliged to wait for the housing's manager to act)).

Furthermore, there is a lack of incentive to undertake this type of anticipatory household adaptations, as the current insurance premium regimes do not reward risk reduction in this way (Nickson et al., 2011; Penning-Rowsell and Priest, 2015). Moreover, publicly visible signs of increased flood protection often maintains or increases anxiety among householders (Harries, 2012), and residents who may wish to sell their homes worry that the adaptations may dissuade future buyers (Harvatta et al., 2011). Transformational adaptations, which are larger actions like moving house to avoid the flood (Kates et al., 2012), are unlikely to be supported by the UK government given the strict allocation criteria for social housing. Overall, this policy appears to fail all three equity criteria, as in order to benefit the onus is placed on the tenant, the social norms to act, their financial ability, and the status of ownership.

Flood insurance

Effective insurance provides a mechanism for quick recovery from the effects of flood damage. The provision of insurance in the UK is based on a unique relationship between the government, which agrees to reduce flood risk, and the insurance companies, which must provide affordable insurance for all (Penning-Rowsell and Priest, 2015). The insurance industry aims to provide the greatest financial return to its shareholders, therefore following the utilitarian model of maximising utility and failing to satisfy the Rawlsian or egalitarian criteria for justice or fairness.

Social housing tenants have been highlighted as a group that infrequently takes out insurance policies, with 29 per cent having no insurance in 2008, and 50 per cent having no contents insurance (vs. 20% on an average income) (Pitt, 2008). With concerns over a lack of affordable insurance with increasing flooding due to climate change, the government and the Association of British Insurers (ABI) have developed the Flood Re scheme using the 2010 Floods and Water Act (JRF, 2014). However, despite the fact that the supposedly affordable insurance premiums for at-risk locations now reflect income levels by them matching council tax bands, insurance only functions when the households take out a policy. Without understanding the barriers to a tenant taking out insurance, and responding accordingly, the Flood Re scheme will have little impact.

Flooding in Hull, UK, and the tenants' capacity to adapt

During summer 2007, the United Kingdom suffered serious floods causing widespread damage and the loss of 13 lives (Pitt, 2008). Local politicians commented after the floods that Hull was a 'forgotten city' in comparison to levels of support for wealthier areas in the UK (BBC, 2007, 1). The floods in Hull illustrate some social justice aspect of FRM policies, based on the experience of residents, written in diaries and documented in reports following the flood.

Hull sits at the intersection of the River Hull and the Humber estuary in the north-east of England. It is liable to both fluvial and tidal flooding, although the floods of June/July 2007 were pluvial (rainfall-related). The rains exceeded the capacity of the drainage system: on 25 June, over 110 mm fell (Whittle et al., 2010), affecting 8,657 households (over 20,000 people) and forcing 31.5 per cent into temporary accommodation (Coulthard et al., 2007a).

This is a deprived part of the UK. Hull in 2007 was the eighth most deprived district in England (Noble et al., 2008). Its residents were above the national average for unemployment (6.2% vs. 3.4% in England), more likely than average to receive welfare benefits (21% vs 14% in England), less likely to have formal working qualifications (40% vs. 29% in England), and more likely to live in rented social housing (28% vs. 13.2% in England) (Whittle et al., 2010; NHS, 2007). These individual indicators do not imply an absolute vulnerability (Wisner et al., 2004). It is the compounding and dynamic effect of these socioeconomic factors in an area that is prone to flooding that makes individuals flood-vulnerable (or prone to suffer what Sayers et al. (2017) term 'flood disadvantage').

Following the flood, Hull City Council (HCC) conducted a widespread survey to determine which properties had been flooded and characterise the affected residents (e.g. by age; income-level). A three-category vulnerability matrix was developed to prioritise action and define HCC's level of support: Gold (residents over 60/with disabilities/single parents with at least one child under 5); Silver (residents without insurance not already categorised as Gold), and Bronze (the remaining residents). Approximately 21 per cent of those affected by flooding lived in local authority-owned homes (which comprise 7% of the total housing stock), of whom 67 per cent were categorised as Gold (Coulthard et al., 2007b). Thus, these social housing tenants were judged disproportionately vulnerable.

The majority of the experiences noted in the diaries of social housing residents concerned property adaptations. The HCC allocated emergency contracts to contractors to undertake the necessary repairs while simultaneously bringing the 48 per cent of stock that was not meeting the Decent Homes standard up to par (Hull City Council, 2008). However, the experiences of both social housing residents and council workers described that work as of an inferior standard, occasionally left incomplete, and taking so long that the repairs lost their priority status and residents were relegated to the normal repairs list, thus prolonging their recovery process (Whittle et al., 2010). This caused feelings of resentment towards those in privately owned accommodation who fared better owing to self-help approaches. Others stated that the Decent Homes objectives were overlooked, causing the contractors to have to return for further building work immediately after they had left.

Many of the social housing residents commented in their diaries that they wished to take charge themselves, to repair and make their homes resilient after the floods, but they felt 'stuck' waiting for the contractors (Whittle et al., 2010, 51). It is also likely that they lacked the authority to help themselves. This feeling of lack of control and the associated stress negatively affecting residents is consistent with other research linking deteriorated mental wellbeing with natural disasters (Harries,

2012). Furthermore, some diary entries noted that they were not consulted and had no control over the repairs and other changes that were undertaken; this lack of tenant consultation is unfortunately typical of much social housing (Shapely, 2010). Despite the Association of British Insurers, a trade lobby organisation, noting that renewing a domestic insurance policy was not a problem, reported renewal quotes increased dramatically after the floods (Coulthard et al., 2007b).

The Hull floods of 2007 illustrate an extreme case of how social deprivation and exposure to flooding can overlap to form spatial clusters of vulnerability. Furthermore, the case study shows that despite being equitable (at least in one definition), the non-structural FRM policies outlined here do not necessarily help if the residents affected lack the capacity or the opportunity (and authority) to adapt so as to cope with the flood risk they face. Subsequent research by Sayers et al. (2017), at a national scale, shows the concentrated nature of geographic flood disadvantage. For example, just ten of the 375 UK local authorities (the ten including Hull) account for 50 per cent of the most socially vulnerable people that live in flood prone areas. The results also highlight the systematic nature of flood disadvantage. For example, flood risks are higher in socially vulnerable communities than elsewhere, particularly in coastal areas (like Hull), dispersed rural communities and economically struggling cities (Hull, again).

Flood insurance and social housing in Scotland

Flood insurance underpins all other FRM strategies in the UK and has been widely available from private insurance companies for many decades (Penning-Rowsell and Priest, 2014). But the propensity for individual residents to purchase domestic household insurance cover, with its flood insurance component, is correlated with the tenure or housing type affected.

The situation in Scotland exemplifies this. Thus those in Scotland living in local authority or housing association properties are unlikely to purchase flood insurance: the average level of deciding not to purchase this cover is over 66.7 per cent (Table 12.1). This is marginally worse than the rest of the UK, but the situation in Scottish private rented furnished accommodation is even more disappointing. Here 80 per cent of tenants made no purchase of insurance for their properties, as reported by the Office of National Statistics (ONS), either for the contents or for the structure (although the latter is not surprising, given that their landlord is responsible for that structure). In contrast, the owners of domestic properties have very low levels of non-insurance, especially those owning the property and paying currently for a mortgage (when such insurance is generally compulsory).

The above situation is exacerbated by the fact that some parts of the rented accommodation sector in Scotland are more prevalent than that south of the border. This particularly applies to local authority property, as in Scotland the rate of occupation of such property is around 50 per cent greater than that in the rest of the UK (i.e. 12.4% compared with 8.5%). However, at least with this property type it is clear that the local authorities are responsible for and are likely to

TABLE 12.1 The lack of take-up of flood insurance by householders in different tenure/housing types

Households not purchasing domestic insurance (tenure type)	Scotland (n=434)	Rest of UK (n=4699)
Local authority	66.7%	63.3%
Housing Association	78.1%	64.4%
Private rental unfurnished	64.5%	60.5%
Private rental furnished	80.0%	78.6%
Owner with mortgage	0.0%	2.1%
Owner outright	1.6%	2.9%
All tenures	22.6%	21.8%

Source: ONS Living Costs and Food Survey, 2014.

Note: 'n' = sample size.

undertake repairs after a flood to the structure of the building – although the Hull example shows some problems here – whereas the private landlords perhaps provide no such guarantee.

What we are seeing here, of course, is a correlation between tenure, income and the propensity to take out flood insurance. Those with low gross weekly incomes tend to live in rented accommodation, particularly local authority rented accommodation, and at the same time often fail to purchase flood insurance for the contents of their homes. One cannot but conclude that to them it is unaffordable. There is evidence for this correlation between tenure and income, such that 66.2 per cent of the poorest households in Scotland live in rented accommodation, whereas 88.9 per cent of the richest 10 per cent of households own their own property, that percentage including both those owning their houses outright and those currently paying for a mortgage.

The situation with regard to rented accommodation and home ownership appears to be getting worse not better. Scottish families, as in the rest of the UK, are now more likely to be renting their accommodation than they were at the peak of owner occupation in 2009, and given the rate of real rises of house prices in the last decade or so – and continuing – this trend is not likely to be temporary.

Conclusions

The findings in Hull, nationally and in Scotland reinforce the need to develop a better understanding of flood risk in socially vulnerable communities if flood risk management efforts are to deliver fair outcomes. This situation raises questions about the production and reproduction of vulnerability (i.e. how vulnerability is sustained). The reproduction of vulnerability could also manifest through increasing exposure to flooding due to climate change (IPCC, 2013; Sayers et al., 2017).

This brings in another social justice aspect, as the poorest 10 per cent of households are responsible for 5 per cent of the UK's household and personal transport greenhouse gas emissions (vs. 16% for the wealthiest 10%) (Hargreaves et al., 2013). Given the most vulnerable to the impacts of climate change in the UK have probably proportionately contributed the least to the creation of the problem there is an ethical argument for providing residents in social housing with much more support for flood risk reduction.

Despite the good intensions of FRM policies to become more equitable and sustainable as a way to reduce flood risk, this chapter shows that for social housing these policies fall short of being socially just (i.e. abiding by the three criteria set out above). The case study in Hull exemplifies how a poor understanding of vulnerability and capacity levels coupled with the existing FRM policies leaves residents feeling unsupported during the flood process. The Scottish ONS data show that many social housing tenants are unprotected by flood insurance helping them to recover from a damaging flood event. The two emerging problems with these non-structural FRM policies are that they require the tenants to, first, voluntarily engage with them, and secondly possess the freedom to do so. Limited personal capacities (e.g. income, education levels, mobility) and constrained freedom to act (e.g. authority for self-help; waiting for a contractor) reduce the ability of residents to undertake flood risk reduction (Porter et al., 2014).

It is therefore clear that more should be done to engage this target group in flood awareness campaigns. It is important to know the barriers to taking out insurance or how residents perceive their autonomy, ability and responsibility for carrying out household adaptations, given the institutional structure of which they are a part. Additionally, households must increase their resilience to flooding at a greater rate than their increasing exposure to flood risk due to climate change. An easy first step would be to increase resilience by mainstreaming some structural adaptations into existing refurbishment programmes (i.e. through linking the Decent Homes Standard to adaptation aims) (Sustainable Homes, 2013).

Social networks also increase resilience to risk (Preston et al., 2014), and neighbourhood groups may be highly effective in raising awareness of flood risk (Terpstra et al., 2009), as they localise the issue and build trust around that knowledge (see chapter 1). Given the institutional nature of housing units (e.g. having wardens, committees), there might already be the social infrastructure in place to build this awareness to increase resilience to flooding. Reports from Hull noted that formally leveraging this social capital to integrate residents and planners to create meaningful emergency plans, while simultaneously sharing knowledge, would reduce future flood risk (Coulthard et al., 2007b) and promote an egalitarian form of justice.

Another way of making FRM policies more socially just for social housing residents would be for the responsible authorities to favour positive discrimination, emphasising the importance of Rawlsian justice. Yet, the challenge with providing the justification for that shift in priority has been exemplified in this paper: vulnerability to risk is not homogeneous. Some residents in social housing lack the capacity to engage with current non-structural FRM policies and require additional support. Yet others in social housing who would also receive that additional support will not face this challenge, causing resources to be ineffective in their allocation. This would be inequitable by every definition of fairness, especially as the resources could have provided support to other residents who may be more vulnerable to flooding (e.g. renters of privately owned dwellings).

There are no simple answers here. However much more effort is needed to understand the multi-faceted dimensions of vulnerability and its differences for residents in all housing tenure types. With this knowledge, more progress may be made in determining the best ways of making FRM interventions more socially just under a changing climate than appears to follow from the current situation.

References

BBC (2007). Flood-hit Hull a 'forgotten city'. BBC News. Available from: http://news.bbc.co.uk/1/hi/england/humber/6270236.stm.

Blair, T. (1998). The third way: New politics for the new century. Fabian Society, London.

Coulthard, T., Frostick, L., Hardcastle, H., Jones, K., Rogers, D., Scott, M. and Bankoff, G. (2007a). The June 2007 floods in Hull: Interim report by the Independent Review Body. Independent Review Body, Kingston-upon-Hull, UK.

Coulthard, T., Frostick, L., Hardcastle, H., Jones, K., Rogers, D., Scott, M. and Bankoff, G. (2007b). The June 2007 floods in Hull: final report by the Independent Review Body. Independent Review Body, Kingston-upon-Hull, UK.

DCLG (Department for Communities and Local Government) (2012). High income social tenants: Pay to Stay consultation paper. Department for Communities and Local Government, London.

Environment Agency (2009). Flooding in England: A national assessment of flood risk. Environment Agency, Bristol.

Environment Agency (2014). Environment Agency corporate plan 2014–2016. EA, Bristol. Available from: www.gov.uk/government/uploads/system/uploads/attachment_data/file/318479/EA_Corporate_Plan_Summary_2014_to_2016.pdf.

Fielding, J. and Burningham, K. (2005) Environmental inequality and flood hazard. *Local Environment*, 10(4), 1–17.

Glynn, S. (2009). *Where the other half lives: Lower-income housing in a neoliberal world.* Pluto Press, London.

Grothmann, T. and Reusswig, F. (2006). People at risk of flooding: Why some residents take precautionary action while others do not. *Natural Hazards*, 38(2), 101–120.

Hargreaves, K., Preston, I., White, V. and Thumim, J. (2013). The distribution of household CO2 emissions in Great Britain. JRF programme paper: Climate change and social justice. Joseph Rowntree Foundation, York.

Harries, T. (2012). The anticipated emotional consequences of adaptive behaviour - impacts on the take-up of household flood-protection measures. *Environment and Planning A*, 44 (3), 649–668.

Harvatta, J., Petts, J. and Chilvers, J. (2011) Understanding householder responses to natural hazards: flooding and sea-level rise comparisons. *Journal of Risk Research*, 14(1), 63–83.

Hollow, M. (2011). Suburban ideals on England's interwar council estates. *Journal of the Garden History Society*, 39(2), 202–217.

Hull City Council (2008). Housing stock assessment. HCC, Hull.

IPCC (2013). *Climate change: The physical science basis: Contribution of Working Group I to the Fifth Assessment Report of the Intergovernmental Panel on Climate Change*, T. F. Stocker et al. (eds). Cambridge University Press, Cambridge.

Johnson, C., Penning-Rowsell, E. C. and Parker, D. J. (2007). Natural and imposed injustices: the challenges in implementing 'fair' flood risk management policy in England. *The Geographical Journal*, 173(4), 374–390.

Johnson, C., Tunstall, S. and Penning-Rowsell, E. C. (2005). Floods as catalysts for policy change: Historical lessons from England and Wales. *International Journal of Water Resources Development*, 21(4), 561–575.

Johnson, C., Tunstall, S., Priest, S., McCarthy, S. W. and Penning-Rowsell, E .C. (2008). Social justice in the context of flood and coastal erosion risk management: A review of policy and practice. Joint Defra/EA Flood and Coastal Erosion Risk Management R&D Programme. Defra, London.

JRF (Joseph Rowntree Foundation) (2014). A briefing on the Water Bill: Flood insurance. Joseph Rowntree Foundation, York.

Kates, R. W., Travis, W. R. and Wilbanks, T. J. (2012). Transformational adaptation when incremental adaptations to climate change are insufficient. *Proceedings of the National Academy of Sciences*, 109(19), 7156–7161.

Lindley, S., O'Neill, J., Kandeh, J., Lawson, N., Christian, R. and O'Neill, M. (2011). Climate change, justice and vulnerability. Joseph Rowntree Foundation, York.

Mill, J. S. (1863). *Utilitarianism*. Dent, London.

NHS (2007). Kingston upon Hull. In Health Profile 2007. Department of Health, London.

Nickson, A., Woolston, H., Daniels, J., Dedring, I., Reid, K., Ranger, K., Clancy, L.Street, R. and Reeder, T. (2011). Managing risks and increasing resilience: The mayor's climate change adaptation strategy. Mayor of London, London.

Noble, M., McLennan, D., Wilkinson, K., Whitworth, A., Barnes, H. and Dibben, C. (2008). The English indices of deprivation 2007. Communities and Local Government, London.

O'Neill, J. and O'Neill, M. (2012). Social justice and the future of flood insurance. Joseph Rowntree Foundation, York.

Penning-Rowsell, E. C. (2015). A realistic assessment of fluvial and coastal flood risk in England and Wales. *Transactions of the Institute of British Geographers*, 40(1), 44–61.

Penning-Rowsell, E. C. and Johnson, C. (2015). The ebb and flow of power: British flood risk management and the politics of scale. *Geoforum*, 62, 131–142.

Penning-Rowsell, E. C. and Priest, S. (2015). Sharing the burden of increasing flood risk: Who pays for flood insurance and flood risk management in the United Kingdom. *Mitigation and Adaptation Strategies for Global Change*, 20(6), 991–1009.

Pitt, M. (2008). Learning lessons from the 2007 floods: An independent review by Sir Michael Pitt. Cabinet Office, London.

Porter, J., Dessai, S. and Tompkins, E. (2014). What do we know about UK household adaptation to climate change? A systematic review. *Climatic Change*, 127, 371–379.

Preston, I., Banks, N., Hargreaves, K., Kazmierczak, A., Lucas, K., Mayne, R., Downing, C. and Street, R. (2014). Climate change and social justice: An evidence review. Joseph Rowntree Foundation, York.

Pringle, P., Paavola, J., Dale, N., Sibille, R., Kent, N. and Le Cornu, E. (2013). PREPARE – Understanding the equity and distributional impacts of climate risks and adaptation options, Part of the PREPARE Programme of research on preparedness, adaptation and risk, Final Report for project ERG1211. Ricardo-AEA for Defra, London.

Rawls, J. (2001) *Justice as fairness: A restatement*. Harvard University Press, Cambridge, MA.

Sayers, P., Penning-Rowsell, E. C. and Horritt, M. (2017). Flood vulnerability, risk and social disadvantage: Current and future patterns in the UK. *Regional Environmental Change*, 18, 339–352.

Sen, A. (1992). *Inequality re-examined*. Clarendon Press, Oxford.

Shapely, P. (2010). Planning, housing and participation in Britain, 1968–1976. *Planning Perspectives*, 26(1), 75–90.

Shelter (2014). Who gets priority for council housing. Shelter, London. Available from: http://england.shelter.org.uk/get_advice/social_housing/applying_for_social_housing/who_gets_priority.

Smit, B. and Wandel, J. (2006). Adaptation, adaptive capacity and vulnerability. *Global Environmental Change*, 6, 282–292.

Sustainable Homes (2013). Your social housing in a changing climate. Climate Change Partnership, London.

Swindon Borough Council (2014). Neighbourhood wardens: A service provided to and paid for by housing tenants and leaseholders. SBC, Swindon. Available from: www.swindon.gov.uk/hs/hs-councilhousing/hs-neighbourhoods-wardens/Pages/hs-councilhousing-wardens.aspx.

Taylor, M. (1998). Combating the social exclusion of housing estates. *Housing Studies*, 13(6), 819–832.

Terpstra, T., Lindell, M. K. and Gutteling, J. M. (2009). Does communicating (flood) risk affect (flood) risk perceptions? Results of a quasi-experimental study. *Risk Analysis*, 29(8), 1141–1155.

The Poverty Site (2014). The poverty site: Key facts. Available from: www.poverty.org.uk/summary/key facts.shtml.

UK Government (2014). Shared ownership for council and housing association tenants. UK Government, London. Available from: www.gov.uk/shared-ownership-tenants.

Whitehead, C. and Scanlon, K. (2007). Social housing in Europe. London School of Economics, London.

Whittle, R., Medd, W., Deeming, H., Kashefi, E., Mort, M., Twigger-Ross, C., Walker, G. and Watson, N. (2010). After the rain: Learning the lessons from flood recovery in Hull. In: Final project report for Flood, Vulnerability and Urban Resilience: A real-time study of local recovery following the floods of June 2007 in Hull. Lancaster University, Lancaster, UK.

Wisner, B., Blaikie, P., Cannon, T. and Davis, I. (2004). *At risk: Natural hazards, people's vulnerability and disasters*. 2nd edition. Routledge, London.

13

EMERGENCY INTENTIONAL FLOODING

Is social justice adequately considered?

Anne Muter

Introduction

When floodwaters are rising, governments should employ existing flood risk management (FRM) plans to limit or control damage. However, under extreme circumstances where flood levels exceed anticipated or planned levels, governments may need to make emergency decisions regarding how to best control them. In some cases, one response in extreme flood conditions is intentionally to direct water to inundate one area for the sake of protecting another, by diverting water or breaching a defence structure (hereinafter 'intentional flooding').

Such intentional flooding in one location may reduce the depth of the flood-water or entirely prevent another area from flooding. While the option of intentional flooding may have direct impacts on the spatial and temporal scope of a flood, it may also affect the distribution of the impacts on different sectors of society and industry, with mixed winners and losers (Penning-Rowsell and Pardoe, 2012).

The Assiniboine River in Canada and the Chao Phraya River in Thailand are in two chronically flood-prone regions, and are used here to demonstrate two recent examples of emergency intentional flooding during severe events. This chapter demonstrates how winners and losers can be chosen by flood managers and asks whether intentional flooding in these cases was socially just and therefore 'fair' (see also chapter 12).

Intentional flooding

Intentional flooding involves the breaching of levees or rerouting of rivers to redistribute flood risk within a basin. Such flooding can provide biodiversity advantages through rewetting and improve geomorphological processes within rivers, but the main objective is generally economic (Goodwell et al., 2014; Popescu et al., 2010; Moloney, 2014). While considerable literature has balanced the hydrological risks and advantages of intentionally breaching levees and rerouting rivers before and during a flood event, it has made limited attempts to weigh the social costs and benefits of these decisions. Indeed, the social justifications used in the literature cited above focus on the financial benefits of protecting one area at the expense of another. An important facet of intentional flooding, when it is carried out outside official management plans, is that unlike pre-agreed flooding or flooding in normal high-risk areas, citizens and land-users are largely unable in emergency situations to build anticipatory resilience to mitigate their flood risk.

Therefore, building on environmental justice literature of allocated hazards (Been, 1993; Thaler et al., 2018) and social justice theories (Bentham and Mill vs. Rawls – see chapter 12), this chapter critiques the impacts of two decisions in Canada and Thailand to implement intentional flooding without pre-event planning.

The Assiniboine River, Canada

The Assiniboine River basin is prone to regular flooding (Bower, 2010), but experienced particularly severe flooding in the summer of 2011. One of the tributaries of the Assiniboine is the Souris River. The Assiniboine meets with the Red River just south of its flow through the City of Winnipeg in Manitoba. All three rivers flooded in 2011. The Souris River's high flow caused the Alameda Reservoir upstream of Winnipeg in Saskatchewan to be at risk of failing. A decision was made to make a small opening in the reservoir wall to reduce the risk of a catastrophic failure (U.S. Army Corps of Engineers, 2012).

The flooding in the Souris River was estimated to be a 1 in 500-year flood and thus exceeded the 100-year design capacity of the Souris basin flood control project (U.S. Army Corps of Engineers, 2012). The Red River was also in a flooded condition as it approached the confluence with the Assiniboine, but it was within range of more normal annual discharges and had been higher within the past decade (CBC, 2011a). At the same time, the flow level in the upper Assiniboine River was very high and threatening to breach the river's dikes. The Assiniboine River flood was estimated to be a 1 in 300-year event (Crabb, 2012).

Two main emergency flood control measures were implemented along the Assiniboine. First, in early May 2011, the Portage Diversion's capacity (a flood control structure) was increased beyond its original design capacity to divert additional excess flows into Lake Manitoba. The Portage Division is a structure that serves to protect the city of Winnipeg. This intervention was done to reduce risk of the Assiniboine breaching its dikes downstream (Manitoba Government, 2013a, b). One perceived impact of this increased diversion by the public was increased flooding of farming and recreational areas around Lake Manitoba (CBC, 2011c). However, in the post-flood report two years later, the government stated that the economic losses around Lake Manitoba would have been just as great in fully unregulated conditions as they were from the 'artificial flooding' related to increasing the diversion (Manitoba Government, 2013a, 8).

Second, a decision was made to breach a dike at the Hoop and Holler Bend in Manitoba to allow flow out of the Assiniboine River in mid-May 2011 (CBC, 2011b). The fear was that if pressure were not released upstream, severe flooding would occur downstream in Winnipeg or a dike breach could occur causing uncontrolled flooding over a 500 km^2 area. The area flooded because of the 'controlled release' was rural farmland used mainly for cattle rearing near the town of Portage la Prairie (Manitoba Government, 2011). The Manitoba government's communication on the intentional flooding included that:

- The controlled release location provided the least risk and best management option for practical implementation reasons;
- The purpose was to minimise the risk of an uncontrolled breach that would affect 850 homes and cover a large area while the controlled release would only impact 150 homes and cover a smaller area;
- The controlled release area was also at some risk from an uncontrolled breach; and, a compensation scheme would be put in place for those in the controlled release area (Manitoba Government, 2011).

The overall impacts of the 2011 Assiniboine River basin floods were historically unprecedented for the region. There was an estimated CAD$1 billion impact on the provincial government's deficit and a government cost of over CAD$1.25 billion (Pursaga, 2012). One of the significant impacts beyond damage to infrastructure and buildings was on agriculture, as millions of acres of land were left unplanted (U.S. Army Corps of Engineers, 2012). The intentional flooding of farmland caused a portion of that farming loss and all the benefit was to the downstream, more urban portions of the Assiniboine catchment, including Winnipeg.

A total of 150 homes and approximately 250 km^2 land was inundated to protect 850 homes at risk from an uncontrolled dike-breach. Homes at high risk of uncontrolled breach were in more urban areas. The value of houses was also likely to have been greater in Winnipeg than in rural areas, thus the cost:benefit approach favoured protecting the higher value of urban property. The provincial government's report on the 2011 floods states that the intentional controlled release saved an estimated CAD$300 million in damage in the lower Assiniboine (Manitoba Government, 2013b). This is consistent with an economic-focused cost:benefit approach to the emergency decision-making.

However, indigenous peoples and communities were seriously impacted by the intentional flooding, including the Lake St. Martin, Dauphin River, Little Saskatchewan and Pinaymootang First Nations. Some 7,100 people were evacuated from their homes and land, of whom 4,000 were Indigenous. As of June 2018, nearly 2,000 First Nation people remained outside of their homes and communities due to flood damages (INAC, 2018). CAD$1.6 million per month was being spent by government on housing and food for those who remain displaced (CBC, 2018). Families had not returned to their communities due to damaged housing, contaminated drinking water and destroyed infrastructure. The displacement of First Nations communities to temporary and unacceptable accommodation caused extensive and profound impacts, including effects relating to health, loss of subsistence and resource livelihoods, loss of community and loss of connection to the land (Thompson et al., 2014).

Several lawsuits were brought against the government for damages caused by the 2011 flood (*Pisclevich v. Manitoba*, para. 4). Four First Nations were successful in certifying a class action lawsuit for compensation for the flooding, including the intentional flooding and displacement from their homes (*Anderson et al. v. Manitoba*). A court approved settlement agreement provides that the Manitoba and Canadian governments are to pay CAD$90 million dollars in compensation to class members from the four First Nations (McKenzie Lake Lawyers, 2018).

The Chao Phraya River, Thailand

Bangkok lies in the Chao Phraya River basin, which experiences annual monsoonal floods (Vojinovic and Abbott, 2012). Such floods impact a large population throughout Thailand and cause significant detrimental economic impacts (Dutta, 2011). In 2011, particularly catastrophic monsoon floods from July to December submerged six million hectares of land and 85 per cent of the country's provinces. The floods were especially economically damaging in the industrial, financial and touristic areas around Bangkok, which contribute 48 per cent of Thailand's gross domestic product (GDP) (Aon Benfield, 2012; Jular, 2017). A series of intense monsoon storms ensured that a flood crisis was maintained throughout the country for five months (Marks, 2015). These floods were one of the costliest recorded in Thai history (US$45.7 billion) (Aon Benfield, 2012; Ziegler et al., 2012).

Major criticism was directed at the Thai government both during and after the flood event. The government took steps to divert waters out of the Chao Phraya River and its tributary canals and storage reservoirs because the swollen waters posed the threat of uncontrolled breaching and inundation of high-value communities and industries in downtown and central Bangkok. Nine billion cubic metres of water were released from dams north of Bangkok in late October 2011. This caused flooding of rural and suburban areas to the west and north of the city, including major industrial estates with global manufacturing significance. The government blocked these surface waters as they approached northern and western Bangkok, preventing them from entering the central and downtown districts by closing flood gates and erecting large sand-bag flood walls (Marks, 2015).

As a result, significant damage occurred in industrial and residential areas, with 73 per cent of Bangkok's low-income residents inundated (Marks, 2015). Almost half the 2,000 industrial estates in the Chao Phraya River basin were flooded and ceased production, permanently or temporarily (Aon Benfield, 2012). Some 300,000 properties in the Bangkok metropole were flooded, with losses to property owners of US$2.7 billion (Aon Benfield, 2012). One impact of the intentional flooding of areas around the downtown locality was that areas that had been modelled to be low flood risk zones were flooded (Koekhumcheng et al., 2012). The flooding of such areas may have exacerbated the impact on people in those areas who were vulnerable due this perceived low risk. To take just one example, a resident in one of the low-risk areas flooded was in denial and failed to take mitigation measures as he relied on his belief that he lived in a low-risk area (Cohen, 2012).

However, due to the diverted floodwaters, the government implemented a mandatory order for 20 per cent of Bangkok's population to evacuate. Of course, only those with sufficient mobility could evacuate, and many ignored formal orders and remained in their neighbourhoods. As the flooding crisis in Bangkok developed, the reported government stance remained as taking all possible measures to protect Bangkok's inner-city commercial and residential districts from flooding (BBC, 2011). But by mid-November, the Thai government gave up on making statements that any area would certainly be protected (Morse, 2011).

Clear animosity was felt towards the government by inundated residents. In northern and eastern Bangkok, residents took direct action against flood measures by sabotaging defences for the downtown area in attempts to relieve the pressure of the water in their own neighbourhoods (Marks, 2015). Some of these residents had water in their neighbourhoods for several weeks (Marks, 2015). Poor, conflicting and false information by the government of their plans to manage water levels aggravated the situation and inhibited households' ability to protect their property. A break in the risk communication chain (see chapters 10 and 11) between government and citizens was demonstrated when satellite data indicated flooding would occur in the Don Muang neighbourhood, but citizens were told there was no flood risk to them. Thus, households in this neighbourhood did not have the opportunity to prepare for the flood in a manner which might otherwise have been possible.

Social justice concerns

There are legitimate utilitarian, egalitarian and Rawlsian social justice approaches; these are more fully explained in chapter 12. The examples in this chapter demonstrate an overwhelming utilitarian, cost:benefit-based approach, but we also consider the benefits that a Rawlsian approach might have delivered to both situations, and the feasibility of implementing such a social justice or 'fairness' principle during emergency intentional flood events.

Canada

The focus on saving more houses at the sacrifice of the few was consistent with a utilitarian approach. But a central criticism of utilitarianism is that it fails to recognise that costs are not borne equally within society. In a 2013 government report on the flood, the high proportion of First Nation evacuees and the First Nations peoples' experience of the flood was recognised as unique amongst Manitobans. The impacts of intentional flooding were disproportionately borne by the Indigenous peoples, inconsistent with egalitarian or Rawlsian justice. Intentional flooding displaced members of seven First Nation communities when the intentional diversion was made to protect urban, recreational and agricultural properties elsewhere. The compensation scheme put in place was the government's only decision made to address the unfairness or inequality of the intentional flooding impacts. Despite their over-representation as an impacted demographic, the 2013 Task Force did not explicitly mention the social justice impacts on Indigenous peoples in its review of events (Manitoba Government, 2013b).

Non-financial considerations of the intentional flooding's impact are also not mentioned in the initial government communication on the scheme or in the post flood report (Manitoba Government, 2011, 2013a, 2013b). Emotional, psychological and environmental implications of the scheme were evident. For example, flooding of a farm may cause permanent damage to a person's or multi-generational family's life work that cannot be compensated by the value of a year's lost crops. The long-term impacts of treating non-Winnipeg Manitobans (rural residents) differently from Winnipeggers (urban residents), and differently still from Indigenous people, could have detrimental effects on overall perceptions of equality in the province.

In several respects, First Nations' relationship with the Provincial government and the natural environment enhanced their vulnerability to the impacts of the deliberate inundation. The unique jurisdictional relationship between the provincial government and First Nations communities delivered a lack of clarity on responsibility to warn and protect communities from floods. Environmentally, First Nations people have a greater reliance upon land-based economies than typical settler communities, where livelihoods, culture and tradition are strongly tied to land. Socio-financially, First Nations communities are typically more deprived than settler communities, with higher rates of poverty and health issues, and therefore have lower resilience to systemic shocks. Therefore, it is very clear that spreading the costs of flooding into such vulnerable communities is not socially just. The capacity for larger, wealthier settler communities in urban areas such as Winnipeg to recover from flood events is likely to have been greater. Although more homes were saved by the decision to breach the Portage Division, reducing the immediate costs of flooding, long-term impacts on vulnerable and potentially dependent Aboriginal groups arguably could outweigh the benefits of spreading on to them the burden of flooding.

Furthermore, the *provincial* taxpayer paid for the government's general compensation package for those intentionally flooded, which for most Manitobans did not translate into a direct benefit. The *urban* residents who were protected and not flooded are likely to have benefitted the most and only paid their small contribution as taxpayers. The compensation scheme was intended to compensate for the losses suffered by those flooded; however, the non-material losses such as emotional stress that are not easily quantifiable (non-pecuniary damages) were still suffered. Many of the intentionally flooded residents remain on the losing end of the scheme, with continuing conflict over the division of reparations amongst differentially affected First Nations communities (CBC, 2018).

A Rawlsian approach to this flood event would have been to ensure maximum protection for the most vulnerable – i.e. the Indigenous communities – and assumed the greater capacity of the higher-value urban settlements to recover quickly from flooding, and in so doing, diverted floodwaters to these areas. However, while this approach would have been more socially just, Been (1993) argues that it would have delivered the path of highest political and legal resistance for government, and would therefore have been unfavourable. In the longer term, Thaler et al. (2018) highlight that supporting the most vulnerable under the Rawlsian framework would encourage moral hazards, encouraging such people to remain on floodplains. However, within the Indigenous setting (and indeed the agricultural community), this argument does not hold, given the strong land-based relationships communities have with an area, which encourage continued occupation of at-risk locations (see also chapter 14).

Bangkok

The city of Bangkok contributes 78 per cent of the Chao Phraya River basin's GDP, containing industry, government and commercial office buildings as the financial core of the country (UN/WAPP, 2003). The areas outside the source of most of this GDP were sacrificed, by intentional flooding, without a focus on benefiting the most disadvantaged groups. The Thai government's prioritisation of Bangkok's protection is therefore consistent with a utilitarian analysis of maximising the total economic protection value. A study prior to this flood event found that Thailand had not considered social justice principles in its flood or water management policies (Lebel, 2007; Lebel et al., 2009). Indeed, President Yingluck demonstrated a single-minded value-orientation, declaring in late November 2011 that the flood protection attempts had been a success since the downtown and inner-city areas had been protected from inundation – this all in the shadow of citizen unrest over the inequity with which the government had dealt with the situation.

In this regard, Marks (2015, 643) quotes one resident of the inundated neighbourhoods as arguing that 'the government was only concerned about impacts to the economy. It did not think about how much people outside the inner city are suffering. And the assistance provided was not enough.' Nonetheless, two of the significant financial losers from the 2011 floods were industry and the insurance business that suffered massive costs from commercial insurance claims (Aon Benfield, 2012). Many of the suburban areas flooded were home to the manufacturing core of Thailand's export economy (Fuller, 2011), which one would have thought a utilitarian approach would have protected with as much vigour as the urban centre. This raises the question as to whether principles other than narrow utilitarian ones contributed to the overall emergency flood management. However, from a social justice perspective, insured industry has perhaps a much greater capacity for recovery of financial losses than middle and low-income homeowners. Therefore, the choice to inundate floodplains where industry was concentrated to protect downstream households could, despite its failure (because in the end the residential areas were flooded), arguably demonstrate an element of Rawlsian thinking.

The focus on downtown Bangkok in the intentional flooding scheme may have also included non-financial social and political motives. The downtown area contains many of the country's culturally, historically, politically and religiously important buildings and monuments, such as the Royal Palace. Marks (2015) argues that the downtown and central areas were so vigorously defended owing to the concentration of elite individuals there, individuals who held significant lobbying capacity in the government for flood protection. However, an egalitarian perspective might consider the cultural value to all citizens of Thailand of the inner city's historical-cultural infrastructure. Perhaps attempting to protect these buildings was done in part on the justification that they comprised a portion of everyone's right of access to the most extensive cultural/historical liberties.

Marks (2015) provides evidence, however, that Rawlsian decision-making *during* the flood event would have been hydrologically impossible. Years of mismanaged urban development, as well as significant obstructions and alterations to the city's drainage system, created a hydraulic system that was fundamentally vulnerable to flood disaster on a large scale. From this perspective, although not socially just in the long term when considering the hydro-institutional management failures throughout Bangkok, in the short term it may have been a shrewd decision to allow what could not be saved to be further sacrificed for the protection of the highest value land. What this highlights, however, is the significant work which must be carried out in the long term to improve Bangkok's infrastructural capacity and land-planning to ensure greater resilience during flood events.

The government's perceived mismanagement of the environmental disaster, combined with the historical antagonism between low-income rural/suburban citizens and elite Thai leaders, exacerbated pressure on the government which eventually fell to a military coup in 2014. Indeed, their method of tackling in the short term of the major and enduringly disruptive flood event may therefore have been short-sighted, with serious longer-term political impacts.

Assessment

These two examples reinforce the conclusion (cf. Johnson et al., 2007) that utilitarian cost:benefit considerations dominate flood risk management and, here, specifically intentional flooding decisions. The Manitoba government much more explicitly disclosed to the public the decision-making considerations than the Thai government. The Manitoba statements and post-flood report were very utility-focused analyses, with little consideration of egalitarian or justice as fairness

principles. The lack of Rawlsian fairness was particularly evident in the disproportionate impact on Indigenous people. On the other hand, the Thai government's
statement to take all necessary measures to protect Bangkok's core did not reveal
explicitly whether social justice played a role in that decision. However, the cultural value of the area that the government chose to protect while allowing
industrial parks to flood, and the political nature of the process, suggests that some
non-financial value based considerations may have been part of the decision-
making processes.

Both examples exhibit instances of short-termism (i.e. thinking about immediate
benefits without consideration of long-term costs). Short-term interventionism has
been criticised by Cooper and McKenna (2008), who argue that the case against
such intervention (e.g. through deliberate inundation) is strongest at the national
and long-term scale. One element of this contention is that the greater scales
include broader considerations such as intergenerational costs and personal responsibility to act, meaning that the costs per benefit are larger at greater scales.
Applying this to the situation of intentional flooding, emergency decision makers
may be failing to include larger-scale costs flowing from broader considerations that
would support a case against intervening by intentionally flooding. As such, it
undermines utilitarian arguments of intentional flooding. For instance, if the long-
term impact of intentional flooding is that a farmer abandons planting in future
years, then the costs to the farming community and the food security of the region
may be far greater than just the short-term cost of one year's lost crops. By disproportionately impacting Canada's First Nations, the longer-term impacts and
perpetuation of historical social inequalities is detrimental to the overall project of
reconciliation between Federal government and Aboriginal peoples (TRC, 2015).

Public engagement is often cited as an effective mechanism for achieving consensus on risk management decisions. However, the nature of emergency flood
decision-making is such that contemporaneous stakeholder participation is very
difficult, if not impossible. However, during advanced planning sessions, stakeholders may be able to inform emergency decisions and improve the social justice
of those decisions. If achievement of social justice includes procedural justice, then
finding a way to include stakeholder participation is a critical mechanism for its
achievement. There is, of course, the counter-argument that introducing ideas of
intentional flooding pre-event may produce such opposition from consulted publics that it eliminates the possibility to implement it as a necessary management
technique during the flood. However, examples from the Mississippi basin in Illi-
nois (Goodwell et al., 2014) demonstrate that reaching consensus at times of crisis is
possible if sufficiently competitive compensatory regimes are available.

The Thai example here showed that, despite attempts to pre-empt and prevent
uncontrolled dike breaches and therefore protect important economic areas of the
city, the intentional flooding measures largely failed, and only one section of the
city was spared. This could indicate that the method of intentional flooding did not
follow an appropriate plan; that the hydraulic system was simply inadequate to
carry the volumes of water passing through the basin; or, as Marks (2015) argues,

inundation decisions were not made based on efficacy, but rather on politics. One must question therefore the long-term socio-political value of deliberately inundating property and infrastructure, when the benefits thereby generated cannot be guaranteed or indeed demonstrated. The communities harbouring floodwaters destined for the downtown area perceived that they had suffered the burden of damage and, in many cases, they in reality had (*Taipei Times*, 2011). There was at least the perception of unfairness and a lack of stakeholder consultation that contributed to the public backlash. The clear divide of winners and losers here was between the rural/suburban populations and downtown urban communities.

Conclusions

A utilitarian cost:benefit approach dominated the decisions to intentionally flood in both instances described here. The maximum number of homes was protected on the Assiniboine in Canada. The greatest GDP producing area – and the prestigious historic core – was the target of flood protection decision-making for the Chao Phraya in Thailand.

Because of those decisions, there was a clear rural/suburban sacrifice for the good of the city in both Canada and Thailand. There was also a focus on economic considerations of trying to prevent the larger economic impact, while harder to quantify non-financial considerations played no obvious role in Canada but may have played some role in Thailand.

Neither scheme met the principle that inequalities should be arranged to most benefit the least advantaged or meet the principle of equal opportunity. Given the huge costs to the general taxpayer and the scale of these disasters, there is significant room for greater consideration of the social justice implications of emergency flood management decisions, difficult though this inevitably might be.

References

Anderson et al. v. Manitoba et al. (2017). MBCA 14. Available at http://canlii.ca/t/gx7zq. [Accessed 28 July 18]

Aon Benfield (2012). 2011 Thailand floods event recap report: Impact forecasting March 2012. Aon Benfield, London.

BBC (2011). Thailand floods: Bangkok 'impossible to protect'. BBC [online], 20 October. www.bbc.co.uk/news/world-asia-pacific-15381227. [Accessed 18 July 18]

Been, V. (1993). What's fairness got to do with it? Environmental justice and the siting of locally undesirable land uses. *Cornell Law Revue*, 78(6).

Bower, S. S. (2010). Natural and unnatural complexities: Flood control along Manitoba's Assiniboine River. *Journal of Historical Geography*, 36(1), 57–67.

CBC (2011a). Dual flood crests could hit Winnipeg. CBC [online], 8 April. Available at www.cbc.ca/news/canada/manitoba/dual-flood-crestscould-   hit-winnipeg-1.1067580. [Accessed 18 July 18]

CBC (2011b). Water flows through Manitoba dike breach. CBC [online], 14 May. Available at www.cbc.ca/news/canada/manitoba/waterflows- through-manitoba-dike-breach-1.975820. [Accessed 18 July 18]

CBC (2011c). Manitoba flood diversion on for Saturday: Premier. CBC [online], 14 May. Available at www.cbc.ca/news/canada/manitoba/manitoba-flood-diversion-on-for-sa turday-premier-1.1073853. [Accessed 18 July 18]

CBC (2018). Judge approves $90M settlement for flooded Manitoba First Nations. CBC [online], 12 January. Available at www.cbc.ca/news/canada/manitoba/manitoba-first-na tions-flooding-settlement-1.4482353. [Accessed 18 July 2018]

Cohen, E. (2012). Flooded: An auto-ethnography of the 2011 Bangkok flood. *ASEAS – Austrian Journal of South-East Asian Studies*, 5(2), 316–334.

Cooper, J. A. G. and McKenna, J. (2008). Social Justice in coastal erosion management: The temporal and spatial dimensions. *Geoforum*, 39, 294–306

Crabb, J. (2012). Brandon residents cope with state of emergency due to flooding. *CTV News* [online], 19 May. Available at https://winnipeg.ctvnews.ca/brandon-residents-cop e-with-state-of-emergency-due-to-flooding-1.642314. [Accessed 28 July 18]

INAC (Indigenous and Northern Affairs Canada) (2018). 2011 Manitoba flood evacuee summary. www.aadnc-aandc.gc.ca/eng/1392047198501/1392047347518. [Accessed 28 July 2018]

Dutta, D. (2011). An integrated tool for assessment of flood vulnerability of coastal cities to sea-level rise and potential socio-economic impacts: a case study in Bangkok, Thailand. *Hydrological Sciences Journal*, 56(5), 805–823.

Fuller, T. (2011). Thailand flooding affects global industries. *The Hindu* [online], 7 November. www.thehindu.com/opinion/op-ed/thailand-flooding-affects-global-indus tries/article2607072.ece. [Accessed 28 July 18]

Goodwell, A. E., Zhu, Z., Dutta, D., Greenberg, J. A., Kumar, P., Garcia, M. H., Rhoads, B. L., Holmes, R. R., Parker, G., Berretta, D. P. and Jacobson, R. B. (2014). Assessment of floodplain vulnerability during extreme Mississippi River flood 2011. *Environmental Science & Technology*, 48(5), 2619–2625.

Johnson, C., Penning-Rowsell, E. C. and Parker, D. J. (2007). Natural and imposed injus-tices: the challenges in implementing 'fair' flood risk management policy in England. *The Geographical Journal*, 173(4), 374–390.

Jular, P. (2017). The 2011 Thailand floods in the Lower Chao Phraya River Basin in Bangkok Metropolis. *Global Water Partnership* [online]. www.gwp.org/globalassets/global/ toolbox/case-studies/asia-and-caucasus/case-study_the-2011-floods-in-chao-phraya-riv er-basin-488.pdf. [Accessed 28 July 18]

Koekhumcheng, Y., Tingsanchali, T. and Clemente, R. S. (2012). Flood risk assessment in the region surrounding the Bangkok Suvarnabhumi Airport. *Water International*, 37(3), 201–217.

Lebel, L. (2007). Adaptation to climate change and social justice: Challenges for flood and disaster management in Thailand. USER Working Paper WP-2007–2012.

Lebel, L., Foran, T.Garden, P. and Manuta, J. B. (2009). Adaptation to climate change and social justice: Challenges for flood and disaster management in Thailand. Chapter 9 in Kabat, O., Ludwig, F. and Van Schaik, H. (eds), *Climate Change Adaptation in the Water Sector*, Earthscan, London.

Manitoba Government (2011). Flood bulletin #41. Manitoba Government website, 10 May. Available at http://news.gov.mb.ca/news/index.html?archive=&item=11454. [Accessed 24 November 2013]

Manitoba Government (2013a). 2011 Flood: Technical review of Lake Manitoba, Lake St. Martin and Assiniboine River water levels. Available at www.gov.mb.ca/mit/floodinfo/ floodproofing/reports/index.html. [Accessed 28 July 18]

Manitoba Government (2013b). Manitoba 2011 Flood Review Task Force: Report to the Minister of Infrastructure and Transportation, April 2013. Available at: www.gov.mb.ca/a sset_library/en/2011flood/flood_review_task_force_report.pdf. [Accessed 24 April 2018]

Marks, D. (2015). The urban political ecology of the 2011 floods in Bangkok: The creation of uneven vulnerabilities. *Pacific Affairs*, 88(3), 623–651.

McKenzie Lake Lawyers (2018). Manitoba Flood Class Action. [Online]. Available at www.mckenzielake.com/practice-areas/class-actions-law/2011-manitoba-flood-class-proceeding. [Accessed 12 April 2018]

Moloney, F. L. (2014). Ethics of artificial levee breaches. *Aquila: The FGCU Student Journal*, 1, 1–7.

Morse, F. (2011). Thailand floods: Government warns that no part of city is safe. *Huffington Post* [online], 27 October. www.huffingtonpost.co.uk/2011/10/27/bangkok-floods_n_1034699.html. [Accessed 28 July 18]

Penning-Rowsell, E. C. and Pardoe, J. (2012). Who loses if flood risk is reduced: should we be concerned? *Area*, 44(2), 152–159.

Pisclevic v. Manitoba (2018). MBQB 52. Available at http://canlii.ca/t/hrm0n. [Accessed 28 July 18]

Popescu, I., Jonoski, A., Van Andel, S. J., Onyari, E. and Moya Quiroga, V. G. (2010). Integrated modelling for flood risk mitigation in Romania: Case study of the Timis–Bega river basin. *International Journal of River Basin Management*, 8(3–4), 269–280.

Pursaga, J. (2012). Manitoba's 2011 Flood Costs could rise even higher. *Winnipeg Sun* [online], 27 November. https://winnipegsun.com/2012/11/27/manitobas-2011-flood-costs-could-rise-even-higher/wcm/53a44eb2-8dcb-4615-a99b-5f3aa6ec3e26. [Accessed 28 July 18]

Taipei Times (2011). Bangkok's neighbours carry flood burden. [Online]. 9 October. www.taipeitimes.com/News/world/archives/2011/10/09/2003515307. [Accessed 18 July 18)

Thaler, T., Zischg, A., Keiler, M. and Fuchs, S. (2018). Allocation of risk and benefits: Distributional justices in mountain hazard management. *Regional Environmental Change*, 18 (2), 353–365.

Thompson, S., Ballard, M. and Martin, D. (2014). Lake St. Martin First Nation community members' experiences of induced displacement: 'We're like refugees'. *Refuge*, 29(2), 75–86

TRC (Truth and Reconciliation Commission of Canada) (2015). Honouring the truth, reconciling for the future: Summary of the final report of the Truth and Reconciliation Commission of Canada. TRC, Winnipeg.

UN/WWAP (United Nations/World Water Assessment Programme) (2003). *1st UN World Water Development Report: Water for People, Water for Life – Pilot Case Studies*. Paris, New York and Oxford: UNESCO and Berghahn Books.

U.S. Army Corps of Engineers (2012). 2011 Post flood report for the Souris River Basin. Water Management and Hydrology Section, St Paul's District.

Vojinovic, Z. and Abbott, M. A. (2012). *Flood risk and social justice*. London:IWA Publishing.

Ziegler, A. D., She, L. H., Tantasarin, C., Jachowski, N. R. and Wasson, R. (2012). Floods, false hope and the future. *Hydrological Processes*, 26(11), 1748–1175.

14

AT THE WATER'S EDGE

Motivations for floodplain occupation

Laura West Fischer

Introduction

Floods are frequent phenomena in the United States of America. Residents across the country are exposed to numerous flood hazards, including fluvial, pluvial, and coastal flooding. Despite the well-documented risk, populations in flood-prone areas continue to increase. In 2015, the Brookings Institution reported that nearly half of all flood loss claims submitted to the Federal Emergency Management Agency (FEMA) since 1978 have come from just 25 of the USA's 3,007 counties (Kane and Puentes, 2015). Since 2000, these same 25 counties have grown by about 2.5 million people, or about 100,000 each, far exceeding the average population growth rate across all US counties of about 15,000. With increased population comes increased development and more assets exposed to flood-related damage and the need to protect both people and property from inundation.

Unfortunately, rising federal flood control expenditure in the United States has not succeeded in reducing flood losses (Parker and Penning-Rowsell, 1983). Post Hurricane Katrina, federal flood expenditure has risen dramatically and is not pro-jected to decrease in the coming decades (Cummins et al., 2010). Protection measures are clearly not reducing the losses associated with flood events. Of the 16 weather and climate events that each caused over $1 billion in damage in the United States in 2017, two were flooding events, three were tropical cyclone events and eight were severe storm events with heavy precipitation that led to flooding (NOAA, 2018).

Various factors including economic activity, water-based amenities and favour-able weather attract people to coastal and riparian areas. As a result, floodplain residents are a heterogeneous group lacking a prototypical occupant. Likewise, various factors, including subsidence, the increased intensity of floods, the increased exposure of people and property to a flood hazard and the increased vulnerability of human populations who live in flood-prone areas, have been put forward to explain the rising costs of floods (Hallegatte et al., 2013). But these theories do not explain why so many different kinds of individuals choose and continue to reside in floodplains despite the well-documented and potentially increasing risk. They do not explain what factors influence the decision to reside in a flood-prone area, the knowledge of which is critical to developing sound flood risk management strategies (Bradbury, 1989).

A matter of perception

Analysis of what motivates people to occupy a flood-prone area invariably involves a discussion of risk perception. Risk perception is defined as the subjective judgments individuals make about the nature and severity of a particular risk. Risk perception researchers, including those who study it within the context of flood hazards, generally frame their analysis within one of two approaches (Birkholz et al., 2014). Borne out of the field of psychology, the rationalist approach focuses on the cognitive models on which individuals rely to make decisions about whether or not to accept a particular risk (Tierney, 1999). By contrast, the sociological or constructivist approach views risk as socially constructed and subject to, among other things, various power structures, organisational dynamics and economic interests (Tierney, 1999) See chapter 1. This chapter argues that, while necessary, framing the decision to reside on a floodplain simply in terms of perception is limiting.

To understand the whole picture, it is necessary to bring in the perspective of environmental justice (EJ) research, which explores patterns of social inequality in relation to a particular environmental amenity or disadvantage. The discipline is concerned with distributional questions such as proximity to an environmental dis-amenity along with procedural questions regarding how these patterns are embedded within existing norms and structures (Walker and Burningham, 2012). Moreover, it provides insight into how an individual's choice decisions can be opened up or constrained by a variety of external factors.

By combining critical insights from these three disciplines (risk perception; constructivist approaches; environmental justice), we suggest in this chapter that the decision to reside in a flood-prone area is not always voluntary. Residents choose to accept the risk of living on a floodplain for myriad reasons, but they may also accept that risk involuntarily. Whether or not the individual makes a voluntary or involuntary decision is dependent on two factors – risk perception and external considerations, which comprise a number of exogenous factors that either open up or constrain an individual's decision-making process. Understood in this way, the question of why someone chooses to reside in a flood-prone area becomes two-pronged. It begins with a question regarding risk perception: why are you living here, despite the risk? It then evolves into an EJ question: what prevents you from moving if you desired to do so?

Two perspectives on risk perception

Bounded rationality, heuristics, and the psychometric paradigm

Early studies of risk perception focused on why individuals deviated from what was considered 'rational' behaviour. Robert Kates, a pioneer of bringing risk perception research into the flood hazard domain, promoted the theory of 'bounded rationality' whereby individuals make decisions about risk within the confines of a specific level of awareness (Kates, 1962). In doing so, their decision does not always align with the most rational choice (Kates, 1962; Slovic, 2000).

According to Kates, individuals are subject to the 'prison of experience' and 'atrophy of time' when deciding whether or not to reside on a floodplain. 'Prisoners of experience' assume that the risk of a flood correlates with the number and type of floods they have experienced. Those who have only experienced minor floods in their lifetimes tend to dismiss the possibility that a major flood could occur in the future. Those befallen by the 'atrophy of time' tend to underestimate the flood risk as their memory of the flood that they have experienced fades with the passage of time.

Subsequent research in the broader field of risk perception laid the foundation for the formulation of heuristics or 'mental strategies' that individuals use to simplify the decision-making process (Slovic, 2000). Particularly relevant to the subfield of hazard risk perception are the availability and affect heuristics articulated by Tversky and Kahneman (1973). The availability heuristic supposes that individuals associate the probability of occurrence of a particular event with the frequency with which they have experienced such an event. The affect heuristic suggests that individuals tend to make judgments about risk based on emotional connections to the hazard (Slovic, 2000).

In the psychometric paradigm, as this school of thought has been labelled, emphasis is placed on predicting behavioural outcomes (Birkholz, 2014). Heuristics, biases and bounded rationality are used to explain when outcomes deviate from what could be considered rational or optimal. This framework confines the determination of risk to individual cognitive processes and posits that risky behaviour stems from the desire to accrue a particular benefit or from bias or from misinformation (Lupton, 1999; Tierney, 1999).

Risk as a social construct

The constructivist perspective, which originates from sociology, rebuts this approach, arguing that individuals do not perceive and make decisions about risk within a vacuum. Moreover, it challenges the notion that risk can be determined with complete objectivity (Lupton, 1999; Hannigan, 2014). The more moderate or 'weak' constructivist perspective acknowledges the hazard as real (see chapter 1), but argues that the risk associated with it can be socially constructed (Hannigan, 2014). The intent is to illuminate societal power dynamics, vested interests, or political, social or economic agendas that might influence the determination or perception of risk (Tierney, 1999). Risk perception is treated as a dependent rather than independent variable (Tierney, 1999). Constructivists argue that one cannot overlook the actors, such as technical experts, policy makers or the media, who are involved in the identification, communication and evaluation of risk and therefore affect its character (Tierney, 1999; Birkholz, 2014).

Within the context of flood risk, much attention has been devoted to how the policy landscape leads to greater encroachment of flood-prone areas. In these cases, 'expert' actors play a significant role in shaping prospective residents' risk perceptions. Parker's (1995) and Pardoe et al.'s (2011) descriptions of the 'escalator effect' helps to explain why we see increased development in the floodplain after structural mitigation measures are implemented to decrease flood risk. Once barriers such as seawalls, dikes, or levees are constructed, communities – feeling overconfident about the safety of the floodplain – may develop it even more, amplifying the flood risk there. These measures, however, can make the floodplain more attractive to residents by 'increasing the perceived safety from the hazard' (Bollens et al., 1988, 311).

McGuire (2015) writes about how policy regimes can create gaps between actual and perceived risk among communities. He demonstrated via the National Flood Insurance Program (NFIP) in the United States that policies that encourage development in hazardous areas could alter individual risk perceptions. Similarly, NRDC (2017) proposed that the NFIP encourages high-risk floodplain occupation by providing homeowners of repetitively flooded properties with incentives to rebuild rather than relocate.

Environmental justice considerations

This chapter argues that risk perception is insufficient in entirely explaining why individuals choose to occupy a flood-prone area. Risk is not distributed evenly nor allocated randomly across society (Walker, 2009). Even if prospective or current residents possess an accurate understanding of flood risk, external factors may constrain their ability to make decisions based on that perception. They may not necessarily reside in a flood-prone area by choice.

Bollens et al. (1988), for example, concluded that occupancy of the floodplain was not due to ignorance of the flood hazard as residents possessed sufficient awareness when deciding where to live. Examples of such external factors include proximity to work (Chan, 1995; Montgomery and Chakraborty, 2015), income level (Willis et al., 2011; Maldonado et al., 2016) and affordability of housing in other areas (Pardoe et al., 2011; Lee and Jung, 2015), cultural connections to the floodplain (Chan, 1995), or policies that discourage relocation (NRDC, 2017). Thus, the social and economic situations in which prospective or current residents find themselves strongly influence how they prioritise flood risk.

Chan (1995), while studying persistent floodplain occupation in eastern Malaysia, found inhabitants not unaware of the flood hazard but inhibited by their low residential and occupational mobility. He also argued that competing incentives facilitated by government subsidies to keep farming led Malaysian farmers to continue to live in flood-prone areas. Lee and Jung (2015) reported the growth of lower-income populations in the riparian floodplains of Austin, Texas, due to low property values inside the floodplain areas. Their research demonstrated that this decrease in property values enticed lower-income families to prioritise affordable housing over living in the floodplain. Pardoe et al. (2011) also demonstrated that the pressure to provide affordable housing leads to encroachment of the floodplain in the United Kingdom.

While previous EJ research related to flooding has explored inequitable exposure to flood risk, Montgomery and Chakraborty (2015) caution against treating floodplains residents as a homogeneous group. To say that only economically disadvantaged populations occupy floodplains would be incorrect. A somewhat ironic but nonetheless important consequence of recent EJ flood-related research has been to challenge the notion that socioeconomically advantaged individuals do not reside in high-risk areas (Chakraborty, 2017; Collins et al., 2017). Amenities shape floodplain occupation just as much as social and economic constraints (Vogt et al., 2008; Montgomery and Chakraborty, 2015; Maldonado et al., 2016).

Willis et al. (2011) surveyed residents of Launceston, Australia and reported that neither lack of awareness of the flood hazard nor miscalculation of the probability and impact of flooding explained why residents chose to live in a flood-prone area. Instead, the decision was largely motivated by a sense of place and a desire to live in a particular area. While the study acknowledges that the poorer residents and tenants of Launceston did feel constrained to live there, others saw it as a benefit that outweighed the risk of flooding (Willis et al., 2011). Vogt et al. (2008) noted that inheriting a family home also encouraged residents to stay. Affluent and middle-income communities may be attracted to live in the floodplain because of beautiful views, proximity to local services, or the spiritual and psychological benefits of residing near water. More affluent residents may also be second homeowners who only inhabit the area for part of the year (Smith et al., 2015). Others, regardless of their socioeconomic status, may choose to continue living there because their families have lived there for generations (Chan, 1995; Willis et al., 2011).

Voluntariness of floodplain occupation

A central question for this topic and its research surrounds the voluntariness of the decision to occupy the floodplain. The co-location of human settlement with natural and human-made hazards is not a new phenomenon, but questions remain about the voluntariness with which people choose to reside near these hazardous areas. In this chapter, voluntariness is understood as the degree to which residents are constrained – by perception or external considerations – in their ability to freely accept the risk of living in a flood-prone area. Best conceived of as a spectrum of ability, a more voluntary decision would be made with less constraint, whereas a less voluntary, or involuntary, decision would be made with more constraint. If the decision is classified as involuntary, it may be reasonable to infer that the flood risk has to some degree been imposed on the resident.

Before proceeding, it is worth mentioning that consideration of the voluntariness of a decision is not meant to assign blame to an external actor or even remove culpability from a resident. Rather, the exercise is meant to deepen understanding of the variety of factors that encourage residency in hazardous areas, namely flood-prone ones. The use of the term 'imposed' in this context is not meant to confer any legal liability, but rather to highlight the lack of choice some floodplain residents face. In other words, the predominant goal is to provide communities and those tasked with mitigating flood risk with insight into the many factors that influence a person's decision to reside on a floodplain. Such insight will hopefully improve discussions about the best way to manage flood risk and improve adaptive capacity.

Factors that influence the decision to live in a flood-prone area can be grouped into two categories – perception and external considerations (Table 14.1). The influence of either category could theoretically lead to a voluntary or involuntary decision. The rationalist perspective on risk perception, for example, might argue that if prospective residents are well informed and understand the biases that originate from heuristics, they could make a more voluntary choice. The constructivist perspective, on the other hand, might argue that the socially constructed nature of risk renders the acceptance of it involuntary (Sjoberg, 1987). For example, policies that encourage development in the floodplain may alter risk perceptions and could lead someone to involuntarily accept a greater degree of flood risk than they would otherwise tolerate.

Environmental justice scholars might argue that particular external considerations like affordability and proximity to work push individuals into involuntary occupation of the floodplain despite their awareness of the risk. Residents who have more flexibility regarding these considerations may be more able to make a voluntary choice. Likewise, residents who choose to prioritise the positive amenities that

TABLE 14.1 Factors that influence the decision to live on a floodplain

Perception		External considerations
Rationalist	*Constructionist/ constructivist*	- Income level - Affordability of housing - Proximity to work - Cultural considerations - Sense of place and familial ties - Policies that incentivise living on a floodplain - Water-based amenities - Spiritual and psychological benefits
- Bounded rationality - Heuristics - Focus on bias and misinformation - Emphasis on predicting behavioural outcomes - Preoccupation with individual cognitive processes	- Risk as a social construct - Risk as a dependent variable - Actors shape identification, communication, and evaluation of risk - Policy landscape shapes risk perception	

living near the water brings when making their decision could also be characterised as making a voluntary choice (assuming they are aware of the flood risk).

Research that attempts to classify the decision to live in a flood-prone area based on this chapter's definition of voluntariness might include surveys of floodplain residents, ethnographic research that illuminates cultural or spiritual connections to the floodplain, or analysis of demographic, socioeconomic and policy trends in the flood-prone area. In one relevant study, Vogt et al. (2008) concluded, via interviews with floodplain residents, that residents were not necessarily engaged in voluntary risk-taking, but rather a process that involved delicate trade-offs (Vogt et al., 2008). Zhang (2010) hypothesised that more informed residents would either voluntarily avoid hazardous locations or apply a mitigation measure to protect their homes. His analysis found that with regard to the flood hazard, risk perception was an important antecedent in the decision to purchase a particular property.

Outside the domain of flood risk, some insight can be drawn from research related to exposure to environmental toxins from polluting plants or hazardous waste treatment, storage and disposal facilities (TSDFs). Environmental justice scholars have debated for decades, with mixed conclusions, about whether or not these facilities are disproportionately located among minority and low-income communities (Mohai and Saha, 2015). Some argue that this phenomenon is due to 'disproportionate siting', while others attribute it to 'minority move-in' (Pastor et al., 2001). Disproportionate siting refers to the perceived disproportional placement locally of undesirable environmental land uses (LULUs) in minority and low-income communities. The key assumption of this hypothesis is that the disadvantaged group lived in the area prior to the siting of the facility. Minority move-in hypothesises that because LULUs decrease neighbourhood quality, lower-income and minority residents will move into the area to take advantage of the lower housing costs. The results of research into this topic are mixed, but there is more evidence in support of the disproportionate siting approach (Mohai and Saha, 2015).

With disproportionate siting, if the community was opposed to the construction of the TSDF, one could argue that the decision to reside in a hazardous area is involuntary because residence predates the existence of the hazard. With minority move-in, the reasoning is more complicated and akin to the context of the flood hazard. In both the cases, involuntary acceptance of risk would result from some external consideration preventing the residents from living elsewhere (i.e. affordability, cultural connections) or adversely influencing their perception of the risk. Moreover, it is hardly reasonable to assume that individuals would voluntarily choose to live near a hazardous area, particularly if doing so exposes them to greater risk without some additional benefit and decreases neighbourhood quality (Zhang, 2010). That being said, unlike with polluting facilities, proximity to the flood hazard may bring benefits, which must be taken into account before determining the voluntariness of a residency decision.

One conclusion of this chapter, therefore, is that holistic frameworks for understanding the diversity of motivations that lead to floodplain occupation are lacking. Some residents living near a hazard have a greater capacity than others to freely accept and respond to the risk associated with that hazard. The measure of this capacity relates directly the voluntariness of their decision to live near that said hazard. It is important for risk managers to be aware of the factors that may influence that capacity, as it will help them better understand the makeup of floodplain residents. Awareness of these motivations may influence the policies they institute to protect residents and manage risk. It would also help them better understand why individuals choose to continue to live in risky areas.

Miami, Florida

It is helpful here to look to an actual floodplain for preliminary evidence of voluntary or involuntary residence.

Miami is located on the southeastern coast of Florida and is statistically the city most likely to be hit by a hurricane in the United States (NOAA, 2014). Due to its coastal location, the metropolitan area is primarily at risk of flooding from storm surge and the heavy precipitation that accompanies hurricanes. The region's topography and porous bedrock increase the risk of inland flooding and its vulnerability to sea level rise. Wahl et al. (2015) report that the risk of compound flooding, which they define as a combination of storm surge and heavy precipitation (see chapter 1), is highest in the Atlantic/Gulf Coast region, where Miami is located.

Ample evidence suggests that Miami is home to a diverse flood-exposed population. As of 2014, it is the eighth largest metropolitan statistical area in the United States and home to a growing population of immigrant and refugee populations (USCB, 2014). Between 2013 and 2014, all three countries in the Miami-Fort Lauderdale-West Palm Beach metropolitan area were among the top 50 counties nationwide that had gained the most population (USCB, 2015). The region also boasts large racial and ethnic diversity, with 65 per cent of the population identifying as Hispanic or Latino (USCB, 2010). Although its scenic coastal views attract affluent homebuyers and sustain a thriving tourist industry, Miami is one of the most unequal cities in the United States in terms of income (Berube and Holmes, 2015).

Multiple studies have reported differences in populations exposed to the two types of flood risk (inland and coastal) in Miami. Chakraborty et al. (2014) and Montgomery and Chakraborty (2015) reported that black and Hispanic residents were significantly more likely to reside in areas exposed to inland flood risk. These areas possessed lower median housing values, lower numbers of vacation homes and less access to the water-based amenities that coastal areas provide (Chakraborty et al., 2014). In other words, they carry many of the costs associated with flood risk, but offer fewer benefits. By contrast, Chakraborty et al. (2014) found that coastal flood risk was associated with higher median household income. Likewise, Montgomery and Chakraborty (2015) found that coastal areas in Miami were associated with less economic insecurity, more vacation homes, and greater access to water-based amenities. Maldonado et al. (2016) found that, in the aggregate in

Miami, higher socioeconomic status was associated with greater flood exposure, likely driven by or reflecting higher value homes.

Once again, this chapter defines voluntariness as the degree to which residents are constrained by perception or external considerations in their ability to freely accept the risk of living in a flood-prone area. In Miami, external considerations, in particular socioeconomic factors linked to coastal amenities, do seem to be influencing the voluntariness with which people accept the risk of living in a flood-prone area. Higher home values in coastal areas may indicate that residents voluntarily accept and expose themselves to flood risk in order to access coastal benefits (Collins et al., 2017). In others words, their decision to live in a flood-prone area could be considered voluntary. Moreover, their higher socioeconomic status and ability to purchase flood insurance enables them to reduce the impact of flood risk. By contrast, individuals living in inland flood zones may not have been able to afford to live in other areas or felt that this was the most reasonable choice for them to make given their financial situation. If this is the case, their decision to live in a flood-prone area could be considered involuntary. Their socioeconomic status constrains their ability to freely accept the risk of living in a flood-prone area and their ability to mitigate that risk.

Another factor identified as influencing risk perception and the voluntariness with which people accept the risk of living in a flood-prone area is the policy landscape. As a reminder, if the policy landscape inaccurately characterises risk and subsequently alters risk perceptions, it may lead a resident involuntarily to accept the risk of living in a flood-prone area. Chakraborty et al. (2014) claim that NFIP subsidisation has created premiums do not reflect actual flood risk and has 'facilitated land speculation, housing development, and residential risk-taking in flood-prone coastal locations' (Chakraborty et al., 2014, 9). Similar to McGuire and NRDC's conclusions presented earlier, these authors posit that federal policies have shaped perception of risk in Miami. This has repercussions for both affluent and lower-income populations living in flood-prone areas since underestimation of risk may lead individuals reasonably to believe that they can afford flood losses and as a result more willingly accept the flood risk that they face. In this case, while the decision to reside in the flood-prone area may be voluntary, the acceptance of the actual flood risk is involuntary.

Interestingly, there is overlap in the literature between studies that explore exposure to flood risk and exposure to environmental toxins in Miami (Collins et al., 2017). That research also usefully compares Miami to Houston, Texas, which contains a diverse population and significant flood risk. Unlike Miami, Houston exhibits more consistent patterns of socioeconomic inequality in relation to flood risk exposure. Largely due to the presence of the petrochemical industry, which removes some of the amenities enticing would-be affluent coastal residents and depresses neighbourhood quality in flood zones, social vulnerability in Houston is positively associated with flood risk and air pollution (Collins et al., 2017). As in inland Miami, if home values in these areas of Houston are depressed due to the flood risk, residents may be involuntarily deciding to live with that flood risk in order to take advantage of the affordable housing.

Conclusions

Flood hazards are numerous and increasing across the United States. As communities and flood risk managers develop strategies to address the growing risk floods pose to people and assets, it is important they understand what motivates individuals to live in flood-prone areas. Doing so will provide them with a more accurate understanding of the factors that directly (i.e. water-based amenities) or indirectly (i.e. housing affordability) lead to greater occupancy of floodplains.

Ultimately, the decision to live in a flood-prone area involves a complex process of risk prioritisation and cost:benefit analysis. However, as this chapter has shown, this decision is not always voluntary. Voluntariness as it is defined here represents the degree to which individuals are constrained by perception or external considerations in their ability to freely accept the risk of living in a flood-prone area. As the description here of toxic discharges from polluting plants and TSDFs demonstrates, the idea of involuntarily accepted environmental risk is not new to the study of environmental hazards, but remains underdeveloped. Both 'minority move-in' and encroachment of flood-prone areas may result from involuntary decisions to occupy risky areas in light of external considerations that take precedence in the decision-making process.

The case of Miami, Florida revealed possible examples of voluntary and involuntary decision-making as it relates to accepting the risk of living in a flood-prone area. In that example it appears that socioeconomic considerations primarily drive the voluntariness of the decision. In order to strengthen these conclusions, future research should consider surveying residents further to confirm that their decision-making process conforms with the concept of voluntariness articulated here.

References

Berube, A. and Holmes, H. (2015). Some cities are still more unequal than others – an update. Brookings Institution. www.brookings.edu/research/reports2/2015/03/city-inequality-berube-holmes.

Birkholz, S., Muro, M., Jeffrey, P. and Smith, H. M. (2014). Rethinking the relationship between flood risk perception and flood management. *Science of the Total Environment*, 478, 12–20.

Bollens, S., Kaiser, E. and Burby, R. (1988). Evaluating the effects of local floodplain management policies on property owner behavior. *Environmental Management*, 12(3), 311–325.

Bradbury, J. A. (1989). The policy implications of differing concepts of risk. *Science, Technology, and Human Values*, 14(4), 380–399. Chakraborty, J., Collins, T. W., Montgomery, M. C. and Grineski, S. E. (2014). Social and spatial inequities in exposure to flood risk: A case study in Miami, Florida. *Natural Hazards Review*, 15(3), 1–10.

Chakraborty, J. (2017). Focus on environmental justice: New directions in international research. *Environmental Research Letters*, 12, 1–4.

Chan, N. W. (1995). Choice and constraints in floodplain occupation: The influence of structural factors on residential location in peninsular Malaysia. *Disasters*, 19(4), 287–307.

Collins, T. W., Grineski, S. E. and Chakraborty, J. (2017). Environmental injustice and flood risk: A conceptual model and case comparison of metropolitan Miami and Houston, USA. *Regional Environmental Change*, 18, 311–323.

Cummins, J. D., Suher, M. and Zanjani, G. (2010). Federal financial exposure to natural catastrophe risk, in: D. Lucas (eds), *Measuring and Managing Federal Financial Risk*. University of Chicago Press, Chicago.

Hallegatte, S., Green, C., Nicholls, R. J. and Corfee-Morlot, J. (2013). Future flood losses in major coastal cities. *Nature Climate Change*, 3(9), 802–806.

Hannigan, J. A. (2014). *Environmental Sociology*. Routledge, London.

Kane, J. and Puentes, R. (2015). In flood-prone areas, a rising tide of population. www.brookings.edu/blog/the-avenue/2015/07/14/in-flood-prone-areas-a-rising-tide-of-population/.

Kates, R. W. (1962). *Hazard and Choice Perception in Flood Plain Management*. University of Chicago Press, Chicago.

Lee, D. and Jung, J. (2015). The growth of low-income population in floodplains: A case study of Austin, TX. *KSCE: Journal of Civil Engineering*, 18(2), 683–693.

Lupton, B. (1999). *Risk*. Taylor & Francis, Abingdon, UK.

Maldonado, A., Collins, T. W., Grineski, S. E. and Chakraborty, J. (2016). Exposure to flood hazards in Miami and Houston: Are Hispanic immigrants at greater risk than other social groups? *International Journal of Environmental Research and Public Health*, 13(8), 775–795.

McGuire, C. J. (2015). U.S. coastal flood insurance, risk perception, and sea-level rise: A perspective. *Coastal Management*, 43(5), 459–464.

Mohai, P. and Soha, R. (2015). Which came first, people or pollution? A review of theory and evidence from longitudinal environmental studies. *Environmental Research Letters*, 10, 1–18.

Montgomery, M. C. and Chakraborty, J. (2015). Assessing the environmental justice consequences of flood risk: a case study in Miami, Florida. *Environmental Research Letters*, 10, 1–24

NOAA (National Oceanographic and Atmospheric Administration) Hurricane Research Division. (2014). What is my chance of being struck by a tropical storm or hurricane? Accessed 31 May 2018, www.aoml.noaa.gov/hrd/tcfaq/G11.html.

NOAA (National Oceanographic and Atmospheric Administration) National Centers for Environmental Information (NCEI) (2018). U.S. billion-dollar weather and climate disasters. Accessed 31 May 2018, www.ncdc.noaa.gov/billions/.

NRDC (2017). Seeking higher ground: How to break the cycle of repeated flooding with climate-smart flood insurance reforms. Accessed May 31, 2018, www.nrdc.org/sites/default/files/climate-smart-flood-insurance-ib.pdf.

Pardoe, J., Penning-Rowsell, E. C. and Tunstall, S. (2011). Floodplain conflicts: Regulation and negotiation. *Natural Hazards and Earth System Science*, 11, 2889–2902.

Parker, D. J. (1995) Floodplain development policy in England and Wales. *Applied Geography*, 15(4), 341–363.

Parker, D. J. and Penning-Rowsell, E. C. (1983). Flood hazard research in Britain. *Progress in Human Geography*, 7(2), 182–202.

Pastor, Jr., M., Sadd, J. and Hipp, J. (2001). Which came first? Toxic facilities, minority move-in, and environmental justice. *Journal of Urban Affairs*, 23(1), 1–21.

Sjoberg, L. (1987). *Risk and Society*, Allen & Unwin, London.

Slovic, P. (2000). *The Perception of Risk*, Earthscan Publications, London.

Smith, A., Newing, A., Quinn, N., Martin, D., Cockings, S. and Neal, J. (2015). Assessing the impact of seasonal population fluctuation on regional flood risk management. *ISPRS International Journal of Geo-Information*, 4(3), 1118–1141.

Tierney, K. J. (1999). Towards a critical sociology of risk. *Sociological Forum*, 14(2), 215–242.

Tversky, A. and Kahneman, D. (1973). Availability: A heuristic for judging frequency and probability. *Cognitive Psychology*, 5, 207–232.

USCB (United States Census Bureau) (2010) American fact finder profile of general population and housing characteristics: 2010 (Miami-Dade County). Accessed 31 May 2018, http://factfinder.census.gov/faces/tableservices/jsf/pages/productview.xhtml?src=CF .

USCB (United States Census Bureau) (2014). American fact finder annual estimates of the resident population: April 1, 2010 to July 1, 2014 – United States – Metropolitan and micropolitan statistical area; and for Puerto Rico. Accessed 31 May 2018, http://factfinder.census.gov/faces/tableservices/jsf/pages/productview.xhtml?src=CF

USCB (United States Census Bureau) (2015). New census bureau population estimates reveal metro areas and counties that propelled growth in Florida and the nation. Accessed 31 May 2018, www.census.gov/newsroom/press-releases/2015/cb15-56.html.

Vogt, M., Willis, K. and Vince, J. (2008). Weighing up the risks: The decision to purchase housing on a floodplain. *The Australian Journal of Emergency Management*, 23(1), 49–53.

Wahl, T., Shaleen, J., Bender, J., Meyers, S. D. and Luther, M. E. (2015). Increasing risk of compound flooding from storm surge and rainfall for major US cities. *Nature Climate Change*, 5(7), 1093–1097.

Walker, G. (2009). Beyond distribution and proximity: Exploring the multiple spatialities of environmental justice. *Antipode*, 41(4), 614–636.

Walker, G. and Burningham, K. (2012). Flood risk, vulnerability, and environmental justice: Evidence and evaluation of inequality in a UK context. *Critical Social Policy*, 31(2), 216–240.

Willis, K. F., Natalier, K. and Revie, M. (2011). Understanding risk, choice, and amenity in an area at risk of flooding. *Housing Studies*, 26(2), 225–239.

Zhang, Y. (2010). Residential housing choice in a multihazard environment: Implications for natural hazards mitigation and community environmental justice. *Journal of Planning Education and Research*, 30(2), 117–131.

15

FLOOD INSURANCE MAPS AND THE US NATIONAL FLOOD INSURANCE PROGRAM

A case for co-production?

Allison Reilly

Introduction

Catastrophic events, like hurricanes Irma, Maria and Harvey in the autumn of 2017, have called into question the efficacy of the US National Flood Insurance Program (NFIP), particularly the program's ability to manage the impacts of climate change. A US Environmental Protection Agency report (2016) concludes that continental US has already experienced increases in frequency of coastal and riparian flooding events. The impacts of climate change paralleled by human population growth and development will make flood-prone communities in the US increasingly vulnerable to flood risks into the twenty-first century (Cleetus, 2014; US Environmental Protection Agency, 2016).

Congressional legislative efforts intended to amend the NFIP have not managed to improve the financial solvency of the program, which has accrued a debt of US $46 billion following the 2017 storms (Gonzalez, 2017). Recent legislative reforms to the NFIP have sought to improve the accuracy of flood insurance rate maps (FIRMs), which are the centrepiece of the program (King, 2013). However, political pressures have restricted Congress in adopting and enforcing FIRMs that reflect the actuarial risk of living in a floodplain (King, 2013). Thus, the NFIPs key risk-communication and decision-making support tool offers a misguided indication of flood risk to communities across the nation (King, 2013; Porter and Demeritt, 2012). Foundationally, the lack of public awareness, understanding, and consequently individual agency over flood risk at the local level is how the NFIP falters.

Through a science and technology studies lens this chapter analyses the implications of how knowledge is currently generated in the FIRM production process, particularly focusing on how different publics are involved. This analysis is premised on the constructionist argument that scientific and social representations of the world directly impact and are impacted by social order, and place limits on how parties can navigate within that order: the ways in which we know and represent the world (both nature and society) are inseparable 'from the ways in which we choose to live in it' (Jasanoff, 2004, 13). Therefore, scientific information, such as flood risk data, is not objective but a form of framing (Chilvers and Kearnes, 2016). More specifically, how risk is framed directly influences the ways in which related governance and solutions are identified and carried out (Lövbrand et al., 2009). Chapter 6 considers this idea from a Dutch knowledge-export perspective. In the case of the NFIP, FIRMs provide an important data-driven element to inform flood risk while simultaneously reflecting the values of the parties that created them (Sismondo, 2010).

This chapter endeavours to analyse critically how FIRMs are generated, who is involved in the production process and what implications the politics of this process have on the outcome of FIRMs and the NFIP more broadly. First, the goals of the NFIP and the lineage of efforts made to improve the program in recent years through FIRM reform are outlined. Then, the Risk MAP procedure – a flood risk map creation project – is compared to an example of co-production in Pickering, England. This chapter then discusses the strengths and weaknesses of these different processes through the lens of Michel Callon's (1999) public participation models. Following this analysis, the chapter will reflect upon and discuss the value of experimental techniques in public participation that could be used to amend and reform the American understanding of flood risk and by default the NFIP.

The National Flood Insurance Program and the role of flood insurance rate maps

Hurricane Betsy, the first billion-dollar natural disaster to hit the US, struck in 1965, serving as a catalyst for Congress to create the National Flood Insurance Program (Dean, 1999). Established in 1968, the NFIP was designed to distribute the burden of flood risks across private homeowners nationwide (Dean, 1999; King, 2013). Managed by the Federal Emergency Management Agency (FEMA), the NFIP provides government administered subsidised flood insurance to US citizens with the goals of reducing flood-related damages and lessening the amount of government funding needed for post-disaster recovery (King, 2013).

Flood insurance rate maps (FIRMs) are the pillar of the NFIP. These maps communicate the degree of flood risk faced by municipalities nationwide (King, 2013). Each FIRM defines the municipality's 100-year floodplain known as the Special Flood Hazard Area (US Federal Emergency Management Agency, 2016a). Along with denoting the Special Flood Hazard Area, FIRMs reflect what is known as the base flood elevation, or the minimum height requirement for properties to be considered flood-proofed (US Federal Emergency Management Agency, 2015). Insurance rates for individual property owners are then generated by comparing the base flood elevation to the height of the first floor of the structure: higher first floors (ground floor in UK parlance) relative to the base flood elevation are associated with lower insurance rates (US Federal Emergency Management Agency, 2013).

By determining the insurance rates, FIRMs dictate how well the NFIP functions to inform pre-flood planning by individuals and communities and to accrue post-disaster funds (Arnell, 2000; King, 2013). While a federal program, the NFIP tries to make use of the authority US municipalities have over land use and planning by encouraging flood-smart practices. Under the NFIP, communities are required to use FIRMs to inform planning and regulations for flood risk management, although they are entitled to adopt more nuanced data sets for decision-making contingent on FEMA approval (US Federal Emergency Management Agency, 2014b). Regardless, city and town planning and hence land use zoning control is integral to the NFIPs performance.

FIRMs underlie the challenges of the NFIP, which has received sharp criticism over the years (King, 2013). After Hurricane Sandy, Congress passed H.R. 41 to increase the borrowing authority of the NFIP to cover the costs of insurance claims made in Northeast US for recovery efforts (King, 2013, 6). Again, in the autumn of 2017, Congress was forced to forgive US$16 billion of the NFIPs debt to avoid the program's financial collapse (Gonzalez, 2017). These recurring instances underscore how the impacts of climate change are continuing to challenge the financial solvency of the NFIP. The two main reasons that the NFIP is financially insolvent are the weak enforcement mechanisms in place to require property owners to purchase insurance policies and the fact that insurance rates do not match the actuarial risk of living in a floodplain (King, 2013; see chapter 3 herein for further discussion of this issue).

Homeowners who reside in a floodplain and have a federally backed mortgage are required under the NFIP to purchase flood insurance, but it is estimated that only 18 per cent of Americans who live in a floodplain are covered under the NFIP (King, 2013, 3). Nationwide, properties vulnerable to the risks of flooding remain uninsured under the NFIP, minimising the funds available to FEMA to use in post-flood scenarios. Second, insurance rates do not reflect the true risk of living in a floodplain, creating a moral hazard along the coast and in at-risk zones (Hecht, 2012, 511). Under the current NFIP system, flood-prone property owners are not required to internalise the actuarial risk that they face (King, 2013).

The federal government has tried to resolve the challenges posed by the NFIP and address its shortcomings in two successive pieces of legislation: the Biggert-Waters Flood Insurance Reform Act of 2012 and the Homeowner Flood Insurance Affordability Act (HFIAA) in 2014. The key provisions of Biggert-Waters were to adjust flood insurance premiums to reflect the actuarial risk of living in a floodplain, require that flood zones reflect the impacts of climate change, and expunge the grandfathering clause that exempted homeowners who built their houses before the inception of the NFIP from purchasing flood insurance. Biggert-Waters initiated a large remapping project, which adjusted FIRMs to reflect the true risk of flooding. However, the annual insurance premiums per property generated from the new FIRMs were incredibly high, jumping for example from $598 per year to $33,000 annually in Union Beach, New Jersey (Palmisano, 2015).

Congress came to the aid of their constituents by passing the HFIAA in 2014 to slow the onset of the insurance premium price hikes. The HFIAA imposes a gradual path to the insurance rates established by the Biggert-Waters mapping project, and re-instates the grandfathering clause. The HFIAA shares the same long-term goal of Biggert-Waters, which is eventually to require homeowners located in the floodplain to purchase flood insurance under the NFIP at rates that reflect the actuarial risk. However, the politically unpopular challenge of increasing insurance rates under the NFIP has been delayed (Palmisano, 2015).

The case for co-production of FIRMs

Congress is struggling to address the cumbersome nature of the NFIP. To date, the efforts put forth have not been able to substantiate a change in the financial solubility of the NFIP nor has FEMA sent a signal of actuarial risk to the public to inspire appropriate proactive flood risk behaviours at the community and individual levels. The underlying challenge facing the NFIP is a widespread lack of awareness and understanding of flood risk (King, 2013). Biggert-Waters set the NFIP on the right course in terms of reassessing FIRMs, but it is not necessarily a revised accuracy of risk maps that will amend the program. Rather, more public participation in the FIRM production process, as a means of capturing multiple types of knowledge, could aid to tackle the challenge of a nationwide misunderstanding of flood risk (Chilvers and Kearnes, 2016).

Currently, the FIRM production process follows what is known as the Risk Mapping, Assessment, and Planning program (Risk MAP) (King, 2013, 14). Adopted as FEMA's national strategy in 2009, Risk MAP aims to provide higher-quality data to vulnerable populations to raise public awareness and improve mitigation strategies (Westcott, 2011). The Risk MAP goal reads, 'Through more precise flood mapping products, risk assessment tools, and planning and outreach support, Risk MAP strengthens local ability to make informed decisions about reducing risk' (US Federal Emergency Management Agency, 2016b). Risk MAP aims to 'engage the community as a partner' by offering multiple points of contact for the communities to engage in the FIRM production process, which is run by federal bureaucrats and engineers (Wescott, 2011, 567).

The process of producing FIRMs consists of a series of meetings between federal staff, local officials, and community members (Table 15.1). To begin the process, local and state officials and community members are invited to a 'discovery meeting' to collect preliminary data and contextual information. The same groups then take part in a 'kick-off meeting' where the Risk MAP project plan is developed, and then a 'flood risk review meeting' during which former flood maps and the flood insurance study are examined. The preliminary FIRM itself, however, is generated separately from these engagements by FEMA's engineers and modellers. Afterwards, community members are welcomed back into the process, but only via a 90-day public comment period, which is mediated through an online platform. Following this period, the FIRM becomes official and the municipality has a six-

TABLE 15.1 Potential barriers to knowledge production within the Risk MAP process

Stage	Potential barrier(s)
Discovery meeting	Community members are invited to attend a meeting led by federal employees. 'Contextual information' is presumably limited to forms that are relevant to traditional scientific data collection.
Kick-off meeting	Federal employees lead the development of the Risk MAP project plan to advance FIRM production for each community.
Flood risk review meeting	Community input is limited to the data solicited in the 'discovery meeting,' and previous FIRM data and the FEMA Flood Insurance Study are prioritised.
FIRM development	FIRMs are developed outside of the proximity of the relevant community and are created independently by FEMA's scientific experts.
Public comment period	Community members are invited back into the process via an online platform where they can provide feedback on the expert-produced FIRMs. This platform is open for 90 days.

month window to comply with the map (US Federal Emergency Management Agency, 2016b).

A potential barrier to generating public awareness of flood risks and in turn improving the efficacy of the NFIP is that the public is kept at arm's length from the risk identification process. The Risk MAP process limits the public's ability to influence the outcome of FIRMs and to develop a comprehensive understanding of flood risk. Community members, while incorporated at the beginning of the FIRM generation process, are later distanced and are only able to influence the final product through an online portal (US Federal Emergency Management Agency, 2014a). The public participation methods associated with the production of FIRMs further reduces the agency that community members have to formalise an understanding of the problem (Callon, 1999). Adjusting these public participation methods has the capacity to open-up new possibilities and understandings of flood risk across the US.

Evidence of the efficacy of alternative public participations models in flood risk management can be found in a case study in Pickering, England: a town with a long history of grappling with riparian flooding events (Landström et al., 2011). In Pickering, a bespoke computer model of floodplain risk was borne out of a competency group. The competency groups, or collaborative in-person meetings, brought together community members and hydrological modellers with experience of flooding in the local area. This group was tasked with redefining the flood risk model in Pickering via discussions facilitated by objects, such as photos, to evoke creative conversations about flood risk management (Landström et al., 2011).

Over the course of many competency group meetings, the community members and the scientific experts equally influenced the kinds of data used in the models, and through this process 'science was mapped into local knowledge' (Landström et al., 2011, 1631). The computer model that captured flood risk in Pickering was a representation borne from the scientific knowledge of experts and the contextual knowledge of residents. This model expressed multiple types of knowledge because of the dynamic nature of the public participation method. In fact, this model served as a platform for Pickering to consider and eventually adopt new land management strategies to tackle flood risk, meanwhile providing deeper awareness of, and greater agency over, flooding for locals (Odoni and Lane, 2010; Landström et al., 2011; Callon, 1999).

Juxtaposing these two models of public participation draws attention to the power dynamics, politics and importance of the role that public participation methods play in identifying and formalising flood risk. Further, it raises questions about what actors are involved and how their knowledge on the topic is regarded in the problem identification. Michel Callon writes extensively on these topics and his theorising neatly assigns the Risk MAP and Pickering scenarios into different categories of public participation (Callon, 1999). Callons theory helps to illuminate the nuanced politics of the different models and most importantly the outcomes that arise from the varying hierarchies of knowledge.

The Risk MAP public participation process emulates Callon's (1999, 87) public debate model. In this model, Callon (1999, 88) argues that inclined publics are given the 'opportunity to speak' in the knowledge production process and the outcomes of the public participation process are deemed legitimate because active publics can be heard. Within the Risk MAP procedures, community knowledge is subsumed into the federally and expertly guided process, maintaining a hierarchy between federal authorities and local actors. This structure creates a demarcation or ranking between federal and scientific knowledge and local knowledge. Throughout the FIRM production process, federal knowledge is thus enriched by secondary local knowledge (Callon, 1999, 86).

Callon (1999, 89) argues that a major shortcoming of this type of participatory knowledge production model is that lay people are still excluded from crafting knowledge that is scientific or traditionally more highly regarded. Local knowledge is less valued than expert knowledge, placing limits on the public's sense of agency and ownership over the process and critically restricts their understanding of flood risk. Publics complement the expert knowledge production process by assessing 'the political, cultural, and ethical implications to certain research,' but do not have agency over constructing their own flood risk 'identity' (Callon, 1999, 87). This participatory knowledge production model, while 'taking into account the existence and diversity of controversial local situations', favours expert knowledge, narrowing the outcome of the process (Landström et al., 2011; Callon, 1999, 88).

The co-production of knowledge model implemented in Pickering, which joined expert and lay participants to collaborate 'on equal footing', allowed active publics to 'gain recognition for their actions' in the knowledge production process

(Callon, 1999, 89, 92). In Pickering, expert knowledge and local knowledge became a 'by-product of a single process', due to the collaborative structure of the competency groups, meaning there was no hierarchy of value or importance between the knowledge generated by experts and that of the public (Callon, 1999, 90; Landström et al., 2011).

Isabelle Stengers (2005) argues for the absolute democratisation of different categories of knowledge. Stengers describes experimental constructivism, which articulates a vision in which vernacular knowledge and experienced knowledge are to be studied with consistency (Stengers, 2005, cited in Whatmore and Landström, 2011, 585). Applying Stengers' philosophy to public participation methods helps to develop an environment in which scientific knowledge is 'put at risk' by knowledge garnered through experience (Stengers, 2005, cited in Whatmore and Landström, 2011, 586). Both Callon and Stengers argue that dismantling the barrier between scientific knowledge and knowledge of the lived experience offers publics the opportunity to 'have a hold over their behaviour ... their identity', thus legitimising the knowledge production process (Callon, 1999, 92; see chapter 6 herein for some challenges in implementing this).

As a result of these adjusted power dynamics between experts and lay people, the outcome of a co-production public participation method for flood risk maps varies from the outcome of a public debate method. In Pickering, co-production transformed the flood risk model from a 'context-less device' to a context-based device (Callon, 1998, cited by Odoni and Lane, 2010, 160). In the public debate model, publics are kept at a distance throughout the process, dropping in their opinions when the opportunity is presented. This distance from the process results in the public's distance from the outcome, failing to offer publics a robust legitimacy of the final product.

Co-production, however, is not a panacea or catch-all solution for flood risk management worldwide. The co-production public participation method has many drawbacks and intricacies that do not allow it to be widely applied. First, co-production is an extremely resource intensive process. At an administrative level, competency groups were crafted by professors and researchers at world-class universities, limiting the availability of such a process universally. At a local level, competency groups were extremely time consuming. Such groups lasted for 12 months, consisting of bimonthly meetings in addition to other intermittent activities (Landström et al, 2011). This time commitment not only limits the kinds of people able to take part in competency groups but also limits the communities where such a model would be viable.

Additionally, the example of success in Pickering is very narrow in scope and cannot necessarily be extrapolated outward to the wide spectrum of municipalities across the United States. Pickering is a small town of only some 6,800 residents in North Yorkshire, England, and is homogenous in regard to ethnic background and language (Welcome to Pickering, 2018). Pickering is a small case study in which competency groups were effective, but this tool might not be transferrable to communities with larger or more heterogeneous populations.

Finally, there is the hindrance of timing. Co-production is most effective and has the potential to garner the most public involvement directly following a flooding event (Whatmore and Landström, 2011). The authors argue that controversy and experience have the capacity to yield active and engaged publics, eager to contribute their context-based understanding of flooding. Flooding is a materially experienced phenomenon and controversies are often found between mainstream scientific knowledge and vernacular knowledge in locations that have experiences with flooding (Whatmore and Landström, 2011, 585). Whatmore (2009, 587) calls these moments of conflict 'knowledge controversies', which are generative events that 'slow down' scientific knowledge that is enmeshed in society, provoking a new awareness of problems and the ways society chooses to approach them. Pickering was a prime location for co-production given its long history with recurrent flooding, but it highlights that not all cities and towns are ideal contexts for co-production methods.

When applied in the appropriate contexts, however, co-production public participation methods have the potential to re-open and reimagine long-held understandings of flood risk. Particularly, a co-production model has the capacity to improve the shortcomings of the Risk MAP public participation process used by the NFIP. Primarily, a co-production model would democratise the power dynamics of the FIRM production process by distributing the power to drive the development of FIRMs amongst all involved parties. The perspectives, experiences, and knowledge of state and local officials, community members and federal employees would be regarded equally in the process of crafting FIRMs (Callon, 1999).

Applying a co-production public participation method to the FIRM production process would also bring community members closer to all aspects of the knowledge production process. Risk MAP shields community members from certain aspects of the process, such as the expert-driven FIRM development itself, and gradually limits community input to an online platform (US Federal Emergency Management Agency, 2016b). A co-production model would help to revitalise the platforms of engagement, closely retaining community members throughout all stages of the process. Additionally, by shifting the power dynamics of the Risk MAP process and removing the barriers for community members to access the platforms of engagement, there would be room for different types of knowledge to influence and shape the outcomes of FIRMs. No longer would traditional scientific knowledge be paramount, but instead scientific knowledge and lived experience would each complement one another (Callon, 1999).

These procedural changes to the FIRM production process have the capacity to improve the flood risk management strategies within the United States NFIP in two key ways. The primary impact is a changed understanding and awareness of flood risk at the community level. Co-produced participatory knowledge production offers a sense of agency to community members since it serves to 'redistribute expertise', equally valuing local voices (Whatmore and Landström, 2011, 586). The secondary impact is the potential for a reimagined depiction of flood risk via flood risk models. Both impacts would allow for improved flood risk management policies that emerge from specific community contexts (Odoni and Lane, 2010).

The longevity of co-production

The lasting impact of co-production, however, is something that ought to be considered outside of the parameters of flood risk management and within the wider context of climate change. More broadly speaking, a co-production public participation process makes known to the public what the knowledge production process looks like. Widely, science has been used as a 'means of public authority' (Wynne, 2006, 212), wielding power while maintaining the public at a distance. A potential positive underlying effect of co-produced participatory knowledge production is a reformed public trust of science (Chilvers and Kearnes, 2016).

Chilvers and Kearnes (2016) argue that public participation is an arbiter for public mistrust of science since the philosophy contests a realist ideology and observes no barrier between objective facts and reason and the areas of culture and values (Jasanoff, 2004, 3). Publics are brought into the fold, and their lived experience is valued equally to expertise (Callon, 1999). In co-production, federal and local actors perform a 'collective investigation of the problem' (Landström et al., 2011, 1619), and produce a visual representation of risk together that can be used as a basis for decision-making. Co-production lifts the veil that currently rests between experts and publics, demystifying the knowledge production process and offering ownership over the process to new participants (Chilvers and Kearnes, 2016; Callon, 1999).

Co-production has the potential to navigate a 'post-truth' political landscape that is now dominant in the United States and in other Western societies (Flood, 2016). Currently, the knowledge-deficit model, or what Callon names 'The Public Education Model' (Callon, 1999, 82), in which experts inform publics about climate science and its impacts, is not effective. Climate change is politically charged in the US, especially following the 2016 election. Co-production offers a means to make climate change a problem that is 'situated' in municipalities nationwide, rather than an intangible global phenomenon (Haraway, 1988, cited by Martello and Jasanoff, 2004, 13; Lövbrand, 2009).

Presently, the United States is combating climate change in two arenas: physically and politically. Revising the NFIP to include a co-production participatory framework could be a means to address both challenges by improving flood risk management strategies and curbing climate scepticism. In an era of post-truth politics, public participation can serve as an 'accountability mechanism' (Chilvers and Kearnes, 2016, 3), offering publics a chance to craft their own climate identity (Callon, 1999). Hopefully, a 'more democratic, more politically accountable, more publicly transparent, more socially responsive' public participation system will yield a more successful relationship between science and democracy (Chilvers and Kearnes, 2016, 3).

Conclusions

This chapter argues that new and experimental public participation methods have the potential to address the underlying problem hindering the efficacy of the NFIP, which is the public's misunderstanding of flood risk nationwide. This science and technologies studies analysis of the NFIP offers potential amendments to the current Risk MAP public participation process, which would ultimately result in a more sophisticated process capable of capturing multiple types of knowledge about flood risk.

Beyond flood risk management, this chapter encourages the use of new and alternative methods of public participation to reimagine understandings of environmental challenges as they are reinvented and transformed globally by climate change. Furthermore, new methods of generating understandings of environmental problems will hopefully inspire a reinvigorated sense of public agency over these issues while simultaneously opening up new opportunities for altered governance and now solutions.

References

Arnell, N. (2000). Chapter 27: Flood insurance. In D. J. Parker (ed.), *Floods*. Routledge, London, 412–424.

Callon, M. (1999). The role of lay people in the production and dissemination of scientific knowledge. *Science, Technology, and Society*, 4(1), 81–94.

Chilvers, J. and Kearnes, M. (2016). Chapter 1: Science, democracy and emergent publics. In Chilvers, J. and Kearnes, M. (eds), *Remaking Participation: Science, Environment and Emergent Publics*. Routledge, London, 1–28.

Cleetus, R. (2014). Overwhelming risk: Rethinking flood insurance in a world of rising seas. Union of Concerned Scientists, Cambridge, MA.

Dean, C. 1999. *Against the Tide: The Battle for America's Beaches*. Columbia University Press, New York.

Flood, A. (2016). Post-truth named the word of the year by Oxford dictionaries. *The Guardian*, 15 November. London.

Gonzalez, G. (2017). House passes disaster relief bill including NFIP debt forgiveness. *Business Insurance*, 13 October, 1.

Hecht, S. (2012). Chapter 14: Insurance. In Fischer Kuh, K. and Gerrard, M. B. (eds), *The Law of Adaptation to Climate Change: U.S. and International Aspects*. ABA Book Publishing, Washington, DC, 511–534.

Jasanoff, S. (2004). *States of Knowledge: The Co-Production of Science and Social Order*. Routledge, London.

King, R. O. (2013). The National Flood Insurance Program: Status and remaining issues for Congress. 42850. Congressional Research Service, Washington, DC.

Landström, C., Whatmore, S. J., Lane, S. N., Odoni, N. A., Ward, N. and Bradley, S. (2011). Coproducing flood risk knowledge: Redistributing expertise in critical participatory modelling. *Environment and Planning A*, 43(7), 1617–1633.

Lövbrand, E., Stripple, J. and Wiman, B. (2009). Earth system governmentality: Reflections on science in the Anthropocene. *Global Environmental Change*, 19, 7–13.

Martello, M. and Jasanoff, S. (2004). Introduction globalization and environmental governance. In M. Martello and S. Jasanoff (eds), *Earthly Politics: Local and Global in Environmental Governance*. MIT Press, Cambridge, MA, and London, 1–30.

Odoni, N. A. and Lane, S. N. (2010). Knowledge-theoretic models in hydrology. *Progress in Physical Geography*, 34(2), 151–171.

Palmisano, J. (2015). American flood insurance: The Biggert-Waters Act and beyond. *Law Street*, 21 February.

Porter, J. and Demeritt, D. (2012). Flood-risk management, mapping, and planning: The institutional politics of decision support in England. *Environment and Planning A*, 44, 2359–2378.

Sismondo, S. (2010). *Expertise and Public Participation: An Introduction to Science and Technology Studies*, 2nd edition. Wiley-Blackwell, West Sussex, UK, 181–204.

Stengers, I. (2005). The cosmopolitical proposal. In Latour, B. and Weibel, P. *Making Things Public: Atmospheres of Democracy*. MIT Press, Cambridge, MA. 994–1003.

US Environmental Protection Agency (2016). Climate change indicators in the United States. EPA 430-R-16-004, 4th edition. US Environmental Protection Agency, Washington, DC.

US Federal Emergency Management Agency (2013). Unit 9: Flood insurance and floodplain management [Homepage of US Department of Homeland Security].

US Federal Emergency Management Agency (2014a). Flood maps: Know your risk and take action against flooding. US Department of Homeland Security, Washington, DC.

US Federal Emergency Management Agency (2014b). Unit 5: The NFIP floodplain management requirements [Homepage of US Department of Homeland Security].

US Federal Emergency Management Agency (2015). Base flood elevation [Homepage of US Department of Homeland Security].US Federal Emergency Management Agency (2016a). Flood zones [Homepage of US Department of Homeland Security], [Online].

US Federal Emergency Management Agency (2016b). What is Risk MAP? [Homepage of US Department of Homeland Security].*Welcome to Pickering* (2018). About the town. [online] www.welcometopickering.co.uk/about-pickering/about-the-town/. Accessed July 28, 2018.

Westcott, J. E. (2011). New Tools from FEMA to help communities understand their coastal risk. Paper delivered at the 2011 Solutions to Coastal Disasters Conference, Anchorage, Alaska, June 25–29.

Whatmore, S. J. (2009). Mapping knowledge controversies: Science, democracy and the redistribution of expertise. *Progress in Human Geography*, 33(5), 587–598.

Whatmore, S. J. and Landström, C. (2011). Flood apprentices: An exercise in making things public. *Economy and Society*, 40(4), 582–610.

Wynne, B. 2006. Public engagement as a means of restoring public trust in science hitting the notes, but missing the music? *Community Genetics*, 9, 211–220.

16

THE EFFECTIVENESS OF SOCIAL MEDIA IN FLOOD RISK COMMUNICATION

Wenhui Wu

Introduction

Flooding is expected to become more frequent and severe under forecast climate change and social-economic development (Haer et al., 2016). There has been, in this context, a gradual change in policy over the last several decades from structural flood defence to flood risk management that emphasises enhancing the resilience of flood-prone communities, in which better risk communication plays a fundamental role (Rollason et al., 2018). The increasing usage of social media (hereinafter referred to as SM) has offered a new channel for this communication, but studies are limited on the evaluation of the effectiveness of this new tool. This chapter examines the use of SM in flood events and proposes four critical factors that determine its effectiveness as a flood risk communication tool. Our aim thereby is to improve the use of SM in the future with the aim of promoting that enhanced resilience.

Flood risk communication

Flood risk communication can take various forms depending on its objectives, ranging from delivering risk information, bringing about behavioural changes to prevent or alleviate flood impacts, enabling public participation in decision-making for flood management plans, and encouraging people to take personal responsibility for managing the risks that they themselves face (Demeritt and Nobert, 2014).

Some organisations, such as the Environment Agency (2015) in the UK, have seen risk communication as focusing on raising the public's awareness of their vulnerability and prompting preventive measures for limiting possible flood impacts. On the other hand, some have considered risk communication to be a continuous process of exchanging information and opinion among stakeholders throughout the flood risk management cycle, covering pre-event warnings, during-event crisis communication and post-event recommendations (Figure 16.1) (Höppner et al., 2012; Lundgren and McMakin, 2013; NOAA, 2016).

We adopt here the latter definition and aims on the basis that in order to build long-term public trust and influence decision-making, agencies and organisations need not only to engage with the public through day-to-day operations before flooding, but they need also to ensure that communication is continuous, if not more frequent, during flood events and afterwards (NOAA, 2016).

Traditional flood risk communication based on a top-down, one-way information dissemination from the authorities to the public has not been successful in improving risk perception and bringing about actions for flood preparedness (Haer et al., 2016). There has been a shift – perhaps as a consequence – from people passively receiving official advice and information to actively seeking information and engaging in communication (Wendling et al., 2013). The importance of social networks and two-way communication in getting messages across and driving actions has therefore been recognised (Haer et al., 2016).

Social media in flood risk communication

Social media can be defined as 'internet-based tools, technologies, and applications that enable interactive communications and content exchange between users who

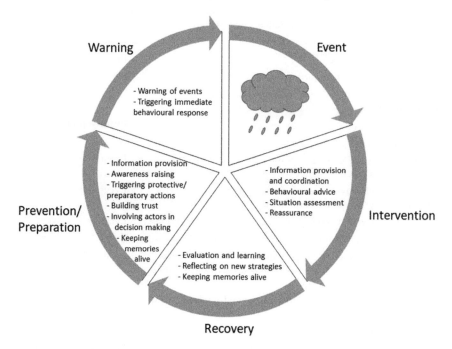

FIGURE 16.1 Risk communication throughout the risk management cycle (after Höpp-ner et al., 2012)

move back and forth easily between roles as content creators and consumers' (Haddow et al., 2017, 174). It includes but is not limited to social networks (Facebook), blogs (WordPress), microblogs (Twitter; Weibo), crowdsourcing (Ushahidi), digital mapping (Google Maps), forums (LiveJournal), content sharing (Youtube; Flikr), and wikis (Haddow et al., 2017; Wendling et al., 2013).

Social media have changed the traditional flood risk communication to an interactive, two-way communication process with multi-dimensional exchange of information and opinions among individuals, communities, governments, media and other organisations (Haddow et al., 2017; Houston et al., 2014). As information circulation becomes decentralised, SM allows rapid dissemination of first-hand information widely among users (Wendling et al., 2013). SM also provides a social context as most connections on SM are between family and friends, as groups within which information is more likely to be processed and acted upon (Lundgren and McMakin, 2013). Thus SM can play various roles in natural disasters such as flooding (Table 16.1) (Houston et al., 2014). The possibility of instant on-site information about flood risk and impact on SM facilitates knowledge sharing and situation awareness building amongst users (Kaewkitipong et al., 2016; Palen and Hughes, 2018). Geographically and temporally traceable data extracted from SM

TABLE 16.1 The use of social media during natural disasters including flooding

Disaster social media use	Disaster phase
Provide and receive disaster preparedness information	Pre-event
Provide and receive disaster warnings	Pre-event
Signal and detect disasters	Pre-event > event
Send and receive requests for help or assistance	Event
Inform others about one's own condition and location and learn about a disaster-affected individual's condition and location	Event
Document and learn what is happening in the disaster	Event > post-event
Deliver and consume news coverage of the disaster	Event > post-event
Provide and receive disaster response information; identify and list ways to assist in the disaster response	Event > post-event
Raise and develop awareness of an event; donate and receive donations; identify and list ways to help or volunteer	Event > post-event
Provide and receive disaster mental/behavioural health support	Event > post-event
Express emotions, concerns, well-wishes; memorialise victims	Event > post-event
Provide and receive information about (and discuss) disaster response, recovery, and rebuilding; tell and hear stories about the disaster	Event > post-event
Discuss socio-political and scientific causes and implications of and responsibility for events	Post-event
(Re)connect community members	Post-event
Implement traditional crisis communication activities	Pre-event > post-event

Source: Haddow et al., 2017.

enables real-time flood detection, monitoring and mapping for better risk communication (Arthur et al., 2018; Restrepo-Estrada et al., 2018; Holderness and
Turpin, 2016; Wendling et al., 2013).

Despite the increasing use of SM, the full potential of SM has yet to be
achieved. As the way people engage in flood risk management changes, communicators should consider more how to make use of SM for even more effective risk
communication. Drawing on examples of SM usage in flood events around the
world, this chapter proposes four factors that we see as determining the effective
use of SM in flood risk communication, in the hope of maximising the benefits of
this relatively new set of tools.

Four factors affecting the effectiveness of social media

Effective use of SM can be defined using different criteria, such as whether actions
are taken after use or how many times a tweet has been forwarded (Lundgren and
McMakin, 2013). In this chapter, SM is considered effective in flood risk communication if its usage leads to the successful passage of information to its intended
audience, resulting in a response to that communication. Its effectiveness depends
on various factors, among which the four proposed here are deemed to be critical.

Access

For risk communication to be successful, messages need to reach the intended
audience. Access refers to the ability of people to obtain the means to use SM, and
reach the information needed. Hence the key question is whether the intended
audience can access SM.

As electronic devices and data plans become more affordable, access to SM
through devices such as laptops, tablets and mobile phones with internet connections has never been so easy. By January 2017 there were already 3.2 billion active
SM users worldwide (Kemp, 2018). The proliferation of SM access leads to its
popularity as a channel for risk communication (Haddow et al., 2017). In the
absence of official information from government agencies, it was SM that provided
timely flood alerts and flood preparedness information generated by active SM
users for Thai citizens during severe Bangkok flooding in 2011(Kaewkitipong et
al., 2016). When Hurricane Sandy hit the US, the Federal Emergency Management Agency was able to disseminate flood-related safety information through SM,
reaching more than 300,000 users on Facebook, and six million Twitter users with
a single tweet (Haddow et al., 2017). Without access to SM, critical information
would not have reached such large numbers in such a short time.

However, it is worth noting that access to SM depends on the resiliency of
electricity and cellular communication infrastructures (Alexander, 2014). With 90
per cent of users accessing SM via their mobile devices (Kemp, 2018), when power
outages last longer than the device's battery capacity, or when the internet disconnects due to failure in infrastructure, access to SM is not available. An example

of such issues comes from the Philippines: as Typhoon Haiyan damaged 90 per cent of power poles and cell-phone connection towers in Tacloban, prolonged power shortage and the breakdown of mobile networks resulted in limited access to SM in the most seriously affected regions, giving it a low rate of effectiveness in communicating flood risk information there (Bichell, 2013).

There is also a social barrier to the access of SM due to digital divides. The extent to which SM is available for use depends on socio-economic, demographic and educational factors (Alexander, 2014). While over half of the world's population has access to the internet, there are still many millions of people without this facility, let alone access to SM (Kemp, 2018). Those without access to SM due to factors such as poverty and disability are also among the most vulnerable to flood risk and the most in need of help through efficient flood risk communication (Alexander, 2014).

Organisational barriers also exist where government authorities face difficulties in obtaining management level endorsement for SM use and operation (Palen and Hughes, 2018). This means that authorities that play key roles in certain stages of the flood risk management cycle such as flood risk identification and recovery may not have official channels for accessing SM to disseminate relevant information to those in need, or receive enquiries and feedback from the public (Stephenson et al., 2017).

Usability

The next question to consider is whether people can successfully use SM to communicate. The usability of SM concerns how easy it is for its users to navigate on SM for access, dissemination and analysis of flood information.

There are some simple but unique features of SM that enables effective communication, one of which is the hashtag on Twitter. A hashtag is composed of a hash symbol (#) followed by any words that categorise the tweet (i.e. message), allowing possible the almost instant search for tweets that share a common topic (Twitter, n.d.). Such a feature is particularly useful in filtering relevant information in flood risk communication (Palen and Hughes, 2018). The hashtag #qldfloods was quickly adopted in Queensland flooding in 2011 by both citizens and government agencies to mark and share flood-related information: more than 35,000 tweets containing the #qldfloods hashtag were generated during the flash floods there, with over a third containing links to further information such as official websites or first-hand photos of the flood (Bruns et al., 2012). During Typhoon Maring in the Philippines in 2014, unified hashtags were suggested, for example #RescuePH, #FloodPH, #ReliefPH, so that people could include these labels in their tweets for more effective searches for information (UNOCHA, 2014).

Another important feature of SM is geotagging. By enabling Global Positioning Systems (GPS) on mobile devices, geotagging allows users to attach locational information to their posts. The provision of real-time geolocation data is vital for the identification of flood occurrence and mapping flood impacts (Restrepo-

Estrada et al., 2018). Such geotagging has also been combined with crowdsourcing for effective flood risk communication (Holderness and Turpin, 2016). For example, during Jakarta's 2014/15 monsoon season, a government-supported crowdsourcing platform called PetaJakarta.org successfully gathered 1,000 geotagged flood reports from Twitter users to present city-scale overviews of the flood situation and impacts for public access and government decision-making (Holderness and Turpin, 2016). It did so by sending automated invitations to users who were tweeting about flooding, in a request to send a geolocated flood verification tweet containing evidence of flood impacts to PetaJakarta.org's Twitter account.

There has been growing interest in 'mining' geolocation data on SM and integrating it with other variables such as precipitation to improve flood warning and mapping (Restrepo-Estrada et al., 2018). However, the use of geotagged data is not without its challenges. With fewer than 1 per cent of tweets being precisely geotagged, for example, researchers have to infer location information from other data on SM such as text content and user profiles (Arthur et al., 2018). Additionally, the sheer amount of locational data can lead to difficulties in extracting and verifying information for effective use (Alexander, 2014; Restrepo-Estrada et al., 2018).

Distinctive design of different SM sites also brings challenges for organisations that lack familiarity with such sites. During the Queensland 2011 floods, the Queensland Police Service (QPS) had arranged for messages on its Facebook page to be automatically crossposted to its Twitter account. However, due to restriction of word counts on Twitter (140 character at the time), lengthy messages from Facebook were not fully displayed on Twitter and hyperlinks became unusable (Bruns et al., 2011). It was not until QPS staff gained better understanding of each sites' characteristics that they decided to use Twitter as the main information channel (Bruns et al., 2011).

The digital divide again places constraints on the usability and hence the effectiveness of SM, as elders and other groups may lack the technological know-how to understand and use SM features effectively (Alexander, 2014). The lack of adequate knowledge in usability of SM among government organisations has also been identified as one of the reasons for the limited endorsement and adaptation of SM (Stephenson et al., 2017).

Trust

Without trust in SM and the information on it, even well prepared and presented information may not lead to effectively communication (Lundgren and McMakin, 2013). Trust here refers to the degree to which the users of SM believe the information they find on their chosen SM is accurate and genuine.

The trust of information on SM is particularly important when SM is the primary information source. In the 2011 Queensland flooding, again, first-hand footages and messages generated by the public on SM were the only means of communication for situational awareness of QPS members in affected areas. But the QPS had built credibility for its social media accounts among citizens through early establishment of its presence, provision of timely and accurate information and active response to enquiries (Queensland Police Service, n.d.). On the basis of this trust, the QPS SM accounts became the primary sources of information for both citizens and mainstream media during the floods (Bruns et al., 2012).

Given the free-to-speak nature of SM, the risk of inaccurate information, rumour and malicious use of SM remains a concern (Houston et al., 2014). The low accuracy of geolocation data on SM, for example, has been noted (Arthur et al., 2018). Rumours such as New York City being flooded after Hurricane Sandy in 2012 were widely spread on SM (Alexander, 2014). Such faulty information could mislead situational awareness, deteriorating the credibility of SM (Lundgren and McMakin, 2013). It also hinders response efforts, as substantial time and resources are needed for verification, leading to delayed action (Stephenson et al., 2017).

On the other hand, studies have found that SM has a certain ability to self-regulate information through collective efforts of officials and volunteers to counter rumours and minimise the spread of misinformation (Palen and Hughes, 2018). For example, during Hurricane Sandy, the tweet claiming the New York Stock Exchange floor had been flooded was quickly debunked by users posting photographs of the flood-free building (Pew Research Centre, 2013). During the 2013 Colorado flooding, the Jefferson County officials also actively addressed rumours regarding the resilience of dams through updates on their official accounts (Kaewkitipong et al., 2016).

Placing constraints on the openness of SM due to caution over misinformation may diminish trust in SM. An example of this is provided by the flash flood in Yuyao, China in 2013, during which the Chinese version of Twitter – Weibo – fell silent as a communication tool, making Yuyao an 'Information Island' (Pang, 2013). Only 170,000 posts were generated about Yuyao's flooding compared to 49.9 million in Lushan's earthquake and 610,000 in Beijing's flooding in 2012 (Pang, 2013), although, admittedly, the scale of impact and public interest in Yuyao was considerably less than that for those other disasters. However, the reduction in Weibo posts appears to have been related to the newly announced regulation on SM by the Chinese authorities, under which 'internet users who make defamatory comments which are visited by 5,000 users or reposted more than 500 times could face up to three years in prison' (BBC, 2013,1). Such tightened government monitoring may have resulted in caution and reluctance among influential Weibo users to disseminate information on the Yuyao flooding and loss of public trust in SM as a channel for open communication (Weng and Ye, 2017).

Another trust-related concern is user privacy and data security on SM. As user-generated data on SM such as geolocation is monitored and mined, it is not clear whether consent for using it has been obtained (Wendling et al., 2013). The majority of people feel they do not have control over their own data on SM and lack confidence in SM companies to protect their privacy (Pew Research Centre, 2018), not least following the revelation in 2018 of unauthorised Facebook information sharing. Public perception of how well their data is protected on SM can influence the extent to which SM is trusted, affecting its usage in flood communication.

Coordination

As SM offers multiple channels of communication and empowers multiple players to participate, communicators need to coordinate both internally and externally for effective communication (Wendling et al., 2013). Coordination involves organisational support for and management of social accounts and collaboration with external partners to achieve wider positive impacts.

Internal coordination within an organisation is necessary to enable successful uptake and smooth operation of its SM accounts. For instance, obtaining senior management support and trust has been identified to be a key to the success of the Incident Management Team at Jefferson County Sheriff's Department (JCIMT) during the 2013 Colorado floods (St. Denis et al., 2014). The JCIMT was given autonomy to proactively communicate with the public on SM without seeking approval from higher authorities (St. Denis et al., 2014). This enabled JCIMT to provide rapid updates and response to the public in the absence of mass media coverage.

Another success of JCIMT was its coordinated operation of accounts across five SM platforms through an integrated communication plan. In the team's own words: 'Twitter is for delivering the news, Facebook is where we talk about the news, and the blog is where we provide the details' (St. Denis et al., 2014, 741). A Google map and a Google form were also deployed to collect public geospatial information and visual evidence of flood damage. All these platforms were interlinked: tweets solicited information for the Google form, which was embedded into the blog containing the map, and posted to Facebook (St. Denis et al., 2014). Not only did such coordination ensure consistency across different platforms, which contributed to trust building, but it also maximised the use of the different SM features to engage with larger groups of users for effective communication.

The multiplicity of players on SM means coordination with external parities is also important (Wendling et al., 2013). The PetaJakarta.org project demonstrated how coordination among the research team at the University of Wollongong, the Jakarta Emergency Management Agency (BPBD DKI Jakarta) and Twitter led to successful crowdsourced flood risk communication between the government and public. The university team developed the platform to match with BPBD's capability and resources and worked with Twitter to generate programmatic invitations and real-time data extraction (Holderness and Turpin, 2016). The project also secured endorsement from Jakarta's governor to call on people to report flooding on PetaJakarta.org. Interaction between the Twitter accounts of the project and BPBD were coordinated to ensure information sharing and corresponding actions (Holderness and Turpin, 2016). By collaborating with trustworthy partners, the project achieved high visibility among both government agencies and citizens, and has been incorporated into BPBD's existing disaster management system (Holderness and Turpin, 2016).

Without coordination, the use of SM could be messy, as evident in the 2011 Thailand floods. At pre-flood preparation stage, flood-relevant government agencies only reported information to the central government, with limited information exchanges between each other (Kaewkitipong et al., 2016). Relying heavily on top-down communication channels before and during early stages of the flooding, the agencies had limited experience with SM (Kaewkitipong et al., 2016). When the agencies finally decided to use SM, working with NGO-initiated crowdsourcing platforms in the midst of the flood, they struggled in allocating sufficient resources to filter and verify information on SM, resulting in sharing inaccurate information with the public (Kaewkitipong et al., 2016). The lack of coordination led to failure in risk communication and loss of public trust.

Hurricane Harvey

Social media played an indispensable role in risk communication during Hurricane Harvey in 2017 such that this devastating storm causing widespread flooding in Houston was termed the US's First Social Media Storm (Rhodan, 2017). SM's widespread uptake by government agencies and citizens had led to its success. Over 120 Houston public agencies had registered on Nextdoor (a neighbourhood-based SM where users' addresses are verified), generating over 4,200 posts to local residents during the hurricane (Lee, 2017). Engagement from the public was also high, with over 10,000 replies and 55,000 'Thanks' in return to public agencies' posts, all of which contributed to a 500 per cent increase in posts and responses on Nextdoor during the event (Sebastian et al., 2017).

Agencies such as the National Weather Service (NWS) and Harris County Sheriff's Office (HCSO) had actively planned for public communication on SM (Lee, 2017; Oyeniyi, 2017). Consistent warnings messages were presented after coordination amongst the NWS and partners, for example precipitation forecast maps on the NWS Twitter page. Coordination among SM officers from major agencies in Harris County also ensured real-time information sharing and misinformation clarification (Oyeniyi, 2017). Innovative features such as geo-targeting on Nextdoor facilitated the HCSO to deliver flood preparation tips to targeted neighbourhoods (Lee, 2017). For the first time, relief effort via SM replaced the traditional 911 emergency telephone lines, as distressed public failed to reach the overloaded emergency call systems and turned to SM for help (Rhodan, 2017).

Despite its success, SM usage also had its shortfalls. First, rumours were not uncommon (Rhodan, 2017; Sebastian et al., 2017). Despite officials' efforts on SM, fake news and inadequate instructions such as staying in a house's attic were still prevalent, causing panic and putting the public's safety at risk (Oyeniyi, 2017; Sebastian et al., 2017). Additionally, citizens had to turn to volunteers on SM due to failure of 911, while government agencies had consistently urged distressed citizens to dial 911 instead of reaching officials on SM (Rhodan, 2017). The agencies did not make use of their established presence on SM for emergency response and failed to coordinate with 911 centres to provide alternatives for SM users. Last but not least, limited collaboration between volunteering groups and government agencies was notable. A closer collaboration with such groups in the future might help agencies filter information more effectively and develop a more comprehensive overview of the situation via coordination across official and volunteer-initiated crowdsourcing platforms (O'Connor, 2017; Sebastian et al., 2017).

Discussion

Although this chapter proposes four factors to be determinants for the effectiveness of SM in risk communication, it is difficult to separate their individual influence on the use of SM in flood risk communication as they are interlinked. Being able to access SM but failing to use its features may limit its effectiveness. Mastering the strengths and limitations of SM is vital for coordination across SM platforms to interact with wider audience and provide more comprehensive information. Trust in an organisation's SM account is affected by proficient use of SM, provision of timely information and active engagement with users, all of which require coordination within the organisation for senior management endorsement and allocation of sufficient resource for timely operation.

Some lessons can be learnt from the above analysis to improve the effectiveness of SM in flood risk communication. First, uninterrupted access of SM can be supported by more robust infrastructure, with back-up power supply and internet connection (Alexander, 2014; Oyeniyi, 2017). For those with access to social media, second, campaigns and training on SM features and methodologies for effective information extraction will help to maximise usability of social media. Third, standardising hashtags by announcing an official hashtag can improve information gathering (UNOCHA, 2014). While the use of functions such as geotagging should be encouraged for better flood warning and monitoring (UNOCHA, 2014), current protection of user privacy is weak and legal advice should be taken by the relevant organisations to generate guidelines to ensure protection of data, without restraining innovative research and applications for SM's usage (Wendling et al., 2013). For those with limited SM access and computer illiteracy, it is important, fourth, that demographic analysis is conducted and alternative access to information is made available to them to ensure those who are most vulnerable to flood impact are not marginalised (Stephenson et al., 2017; UNOCHA, 2014).

More than that, coordinated communication with the public on SM throughout the flood risk management cycle is also needed. Uptake of SM among flood management-related government authorities remains relatively low compared with the public; it is therefore important to encourage organisational endorsement and support for SM use across related organisations during different stages of flood risk communication (Stephenson 2017). For those with access to SM, establishing a presence on SM well before a flood and raising public awareness of official accounts on SM are vital so that people know where to look for information and what sources to trust on SM when faced with a sudden flood event (NOAA, 2016). Timely and active responses to public enquiries and smooth operation across SM platforms are necessary for further strengthening relationships with the community (Lundgren and McMakin, 2013). When dealing with misinformation, instead of putting constraints on the openness of SM, proactively issuing official information directly through reliable channels and steering conversation to clarify information will be more effective (Environmental Agency, 2015; Lundgren and McMakin, 2013). Collaboration with trusted partners on SM is among the best practices for risk communication as it enhances the communicators' ability to process information and to reach wider audiences via different channels to drive appropriate and widespread dislocation minimising action (NOAA, 2016; O'Connor, 2017)

Finally, a comprehensive risk communication strategy capturing the above-mentioned practices is needed to set guidance for SM usage in effective flood risk communication. It will enable communicators to adopt and operate SM in an organised manner, identify target audiences and working partners, and maximise the use of available resources and reach SM's full potential.

Conclusions

This chapter has identified four factors that determine the effectiveness of SM in flood risk communication: access, usability, trust and coordination. They are fundamental to the physical and social adaptation and operation of SM. While they are critical to the success of SM usage, they do not cover every aspect as the response to flood risk via communication depends on a range of complex social, psychological and situational factors (NOAA, 2016). The content and format of information on SM can play an important role in attracting public attention and driving actions (Rollason et al., 2018). The style of communication chosen by organisations to engage with citizens on SM also has significant impact on public perceptions. The heterogeneity of the public means a people-oriented SM communication strategy based on the diverse needs of target audiences will be more effective in driving attitude and behavioural changes (Haer et al., 2016; Ping et al., 2016).

Although this chapter has attempted to evaluate the effectiveness of social media based on criteria such as the scale of the audience reached and online response, it is difficult to identify how many within intended audiences have received the information they needed, and even more so to quantify the extent to which SM has positively influenced risk perceptions and behavioural change, which is a fundamental objective of risk communication for many government organisations (Wendling et al., 2013). So inevitably this assessment remains incomplete.

Nevertheless, given the continuous growth in public uptake, SM is expected in the future to see an even more widespread and proficient use among government agencies (Alexander, 2014). Multidisciplinary research both in depth and with breadth is needed to further explore best practices for social media in flood risk management. Such future research should focus on developing better mechanisms for real-time rumour detection, data extraction and fusion with other data for applications throughout the flood risk management cycle. Coordination among different organisations across different SM platforms will enable more comprehensively covered flood risk communication to reach wider audiences and thereby promote their greater flood resilience.

References

Alexander, D. E. (2014). Social media in disaster risk reduction and crisis management, *Science and Engineering Ethics*, 20(3), 717–733.

Arthur, R., Boulton, C. A., Shotton, H. and Williams, H. T. P. (2018). Social sensing of floods in the UK. *PLoS ONE*, 13(1). https://doi.org/10.1371/journal.pone.0189327 e0189327.

BBC (2013). China issues new internet rules that include jail time. *BBC*, 9 September, www.bbc.com/news/world-asia-china-23990674.

Bichell, R. E. (2013). Flooded and powerless: When lights and cellphones go dark. *National Public Radio*, 13 November, www.npr.org/sections/alltechconsidered/2013/11/12/244796515/flooded-and-powerless-when-lights-and-cell-phones-go-dark.

Bruns, A., Burgess, J., Crawford, K. and Shaw, F. (2012). # qldfloods and@ QPSMedia: Crisis communication on Twitter in the 2011 south east Queensland floods. ARC Centre of Excellence for Creative Industries and Innovation, Brisbane.

Demeritt, D. and Nobert, S. (2014). Models of best practice in flood risk communication and management. *Environmental Hazards*, 13(4), 313–328 .

Environmental Agency (2015). Public dialogues on flood risk communication: Literature review. Environment Agency, Bristol.

Haddow, G. D., Bullock, J. A. and Coppola, D. P. (2017). *Introduction to Emergency Management*, 6th edition. London: Butterworth-Heinemann.

Haer, T., Botzen, W. J. W. and Aerts, J. C. J. H. (2016). The effectiveness of flood risk communication strategies and the influence of social networks: Insights from an agent-based model. *Environmental Science and Policy*, 60, 44–52.

Holderness, T. and Turpin, E. (2016). From social media to geosocial intelligence: Crowd-sourcing civic co-management for flood response in Jakarta, Indonesia. In *Social Media for Government Services*. New York: Springer.

Höppner, C.Whittle, R., Bründl, M. and Buchecker, M. (2012). Linking social capacities and risk communication in Europe: A gap between theory and practice? *Natural Hazards*, 64, 1753–1778.

Houston, J. B., Hawthorne, J., Perreault, M. F., Park, E. H., Hode, M. G., Halliwell, M. R., McGowen, E. T., Davis, R., Vaid, S., McElderry, J. A. and Griffith, S. A. (2014). Social media and disasters: A functional framework for social media use in disaster planning , response , and research. *Disasters*, 39(1), 1–22.

Kaewkitipong, L., Chen, C. C. and Ractham, P. (2016). A community-based approach to sharing knowledge before, during, and after crisis events: A case study from Thailand. *Computers in Human Behavior*, 54, 653–666.

Kemp, S. (2018). Digital in 2018: World's internet users pass the 4 billion mark. https://wearesocial.com /uk/blog/2018/01/global-digital-report-2018.

Lee, C. (2017). How social media assisted cops with the Hurricane Harvey response. *Policeone*, 11 December. www.policeone.com/police-products/emergency-preparedness/articles/467431006-How-social-media-assisted-cops-with-the-Hurricane-Harvey-response/.

Lundgren, R. E. and McMakin, A. H. (2013). *Risk Communication: A Handbook for Communicating Environmental, Safety, and Health Risks*, 5th edition. Hoboken, NJ: John Wiley & Sons.

NOAA (National Oceanic Atmospheric Administration) (2016). Risk communication and behavior: Best practices and research findings. NOAA Social Science Committee, Silver Spring, USA.

O'Connor, L. (2017). How we could better leverage social media during disasters like Harvey. *HuffingtonPost*, 6 September. www.huffingtonpost.com/entry/harvey-social-media_us_59a47f42e4b050afa90c1049.

Oyeniyi, D. (2017). How Hurricane Harvey changed social media disaster relief. *TexasMonthly*, 8 October. www.texasmonthly.com/the-daily-post/how-social-media-managers-responded-to-hurricane-harvey/.

Palen, L. and Hughes, A. L. (2018). Social media in disaster communication. In Rodríguez, H., Donner, W. and Trainor, J. (eds), *Handbook of Disaster Research. Handbooks of Sociology and Social Research*. New York: Springer.

Pang, H. (2013). *Hurui PANG: Weekly public opinion observation: china's Internet we-media information supply in Yuyao flooding is significantly less than that of Ya'an earthquake* (original title in Chinese庞胡瑞囚余姚水灾中国网络自媒体信息供给远低于雅安地震-一周舆情-人民网), *People.cn*, 14 October, http://yuqing.people.com.cn/n/2013/1014/c364176-23189976.html.

Pew Research Centre (2013). Twitter served as a lifeline of information during Hurricane Sandy. *Pew Research Centre*, 28 October. www.pewresearch.org/fact-tank/2013/10/28/twitter-served-as-a-lifeline-of-information-during-hurricane-sandy/.

Pew Research Centre (2018). How Americans feel about social media and privacy, *Pew Research Centre*, 27 March. www.pewresearch.org/fact-tank/2018/03/27/americans-complicated-feelings-about-social-media-in-an-era-of-privacy-concerns/.

Ping, N. S., Wehn, U., Zevenbergen, C. and van der Zaag, P. (2016). Towards two-way flood risk communication: Current practice in a community in the UK. *Journal of Water and Climate Change*, 7(4), 651–664.

Queensland Police Service (no date). Queensland Police Service disaster management and social media: A case study. www.police.qld.gov.au/corporatedocs/reportsPublications/other/Documents/QPSSocialMediaCaseStudy.pdf.

Restrepo-Estrada, C., de Andrade, S. C., Abe, N., Fava, M. C., Mendiondo, E. M., de Albuquerque, J. P. (2018). Geo-social media as a proxy for hydrometeorological data for streamflow estimation and to improve flood monitoring. *Computers and Geosciences*, 111, 148–158.

Rhodan, M. (2017). Hurricane Harvey: The U.S.'s first social media storm. *TIME*, 30 August. http://time.com/4921961/hurricane-harvey-twitter-facebook-social-media/.

Rollason, E., Bracker, L. J., Hardy, R. J., Large, A. R. G. (2018). Rethinking flood risk communication, *Natural Hazards*, 92(3), 1665–1686.

Sebastian, T., Lendering, K., Kothuis, B., Brand, N., Jonkman, B., van Gelder, P., Godfroij, M., Kolen, B., Comes, T. and Lhermitte, S. (2017). Hurricane Harvey Report: A fact-finding effort in the direct aftermath of Hurricane Harvey in the Greater Houston Region. 19 October. Delft University Publishers, Delft.

St. Denis, L. A., Anderson, K. M. and Palen, L. (2014). Mastering social media: An analysis of Jefferson County's communications during the 2013 Colorado floods. In *Proceedings of the 11th International Iscram Conference*, 737–746.

Stephenson, J., Vaganay, M., Coon, D., Cameron, R. and Hewitt, N. (2017). The role of Facebook and Twitter as organisational communication platforms in relation to flood events in Northern Ireland. *Journal of Flood Risk Management*. https://doi.org/10.1111/jfr3.12329.

Twitter (no date). How to use hashtags. https://help.twitter.com/en/using-twitter/how-to-use-hashtags.

UNOCHA (2014). Hashtag standards for emergencies. OCHA Policy and Studies Series, United Nations Office for the Coordination of Humanitarian Affairs. www.unocha.org/sites/unocha/files/ Hashtag Standards for Emergencies.pdf.

Wendling, C., Radisch, J. and Jacobzone, S. (2013). The use of social media in risk and crisis communication. OECD Working Papers on Public Governance, No. 24. OECD Publishing, Paris.

Weng, S. and Ye, X. (2017). Risk society and effective public governance: Analysis on City Y's flooding, *Chinese Public Administration*, 8, 145–149 (original article in Chinese: 翁士洪 & 叶笑云 (2017) 风险社会与有效公共治理:对Y市水灾的社会剖析，中国行政管理⊠第8期⊠145–149页).

INDEX

accountability 7–10, 12, 20, 51, 185
actor mapping 43–8, 50, 53, 55
actor-network theory 44, 47
adaptation 43–57, 71, 91–102, 115, 125, 141–52, 170, 193, 198
Advocacy Coalition Framework (ACF) 79, 94, 99–100, 103–12; and deep core beliefs 104–7, 112; and secondary beliefs 104
Afghanistan 122, 125
Africa 71, 118–19, 132
agency/agencies /agents 1–6, 9, 12, 26, 44, 47, 50–3, 98–99; and coastal floods 55; and flood insurance maps 177, 181–2, 184, 186; and policy belief change 104–6, 108, 110–11; and policy mobilisation 70; and power shifts 59–60; and social media 189, 191–2, 194, 196, 198; and warning systems 117–21, 123, 126, 133, 138
agenda-setting 4, 17, 51, 60, 66, 80–1, 84, 86–8, 91, 94–8, 124, 168
Agosto Filion, N. 96–7
Alameda Reservoir 154
Albrecht, J. 20
Albright, E. 104–5
alienation 9
Amazon River 119
Argentina 79
Aristotle 59
Arnell, N. 33–6
Arroyo, V. 97

Asia 81, 118–19, 125; and the Babai River 126; and Central Asia 118; and the Kamala River 126
assemblages 70, 74
Assiniboine River 153–5, 162
Association of British Insurers (ABI) 145, 147
Atlantic 172
Australia 17, 116, 118, 123, 169
autonomy 19, 47, 51, 64, 144, 149, 195
availability heuristic 123
Aviva 33, 35

Bachrach, P. 44, 49
Bacon, F. 5
Bakker, H. 119
Baldi, B. 66
Baldini, G. 66
Bangladesh 115, 119–25
Bankhoff, G. 5
Banks, H.O. 106
Baratz, M.S. 44, 49
Baumgartner, F.R. 93
Becker, M. 1–13, 17–26
Been, V. 159
behavioural approach 60
Bentham, J. 154
Beven, K. 6
Bhutan 122, 124–5
bias 6–7, 46, 50, 62, 105, 109, 167, 170
biology 8, 12, 109

Birkland, T.A. 85
Bollens, S. 168
Bond, H. 30–42
bounded rationality 167, 170
Brahmaputra River 123
bureaucracy 120, 124, 132, 180
Butler, C. 9, 19

California 7, 103–14; and the CALFEED
   Bay Delta program 111; and the
   California Water Plan 106–11; and the
   Clean Water Act 107; and Department of
   Water Resources (DWR) 104, 106–11;
   and the Los Angeles River 105
Callon, M. 7, 25, 178, 182–3, 185
Cambodia 115, 119
Cameron, C. 51
Canada 5, 9, 17, 30–42, 79, 153–4, 156–9,
   161–2; and the Little Saskatchewan First
   Nation communities 155; and the
   National Disaster Mitigation Program 34;
   and the Pinaymootang First Nation
   communities 155
canals 156
capability 122, 125, 142–3
capacity 12, 59, 63, 73, 76; and emergency
   intentional flooding 154, 158–60; and
   flood insurance maps 181, 184; and
   floodplain occupation 170, 172; and
   Multiple Streams model 84, 91; and
   social housing 141–3, 145–7, 149; and
   social media 191; and warning systems
   117–18, 126–7, 132, 135–6
Caribbean 4, 43–57, 94–5; and the
   Caribbean Climate Change Centre
   (CCCC) 47, 50; and Caribbean Small
   Island Developing States (CSIDS) 43–57
catalytic change 93, 95, 99–100, 178
catchment planning 63–4
Central Valley Project Improvement Act
   108
centralisation 21, 59, 61–2, 65–6, 103, 117,
   120–2, 125–6, 137; and bargained
   centralism 61–2, 65; strong centralisation
   62
Chakraborty, J. 169, 172–3
challenges 1–5, 7, 10–13, 17, 19; and
   finance 31, 39; and flood insurance maps
   179–80, 183, 185–6; and floodplain
   occupation 167, 169; and social housing
   149; and social media 193; and warning
   systems 115–30, 132, 136, 138
Chan, N.W. 168
Chilvers, J. 185
China 7, 115, 120, 124, 194

class 143
climate change 13, 18, 23, 26, 30–1; and
   flood insurance maps 177, 179, 185–6;
   and policy evolution 91–102; and policy
   mobilisation 71; and power for change
   43, 51, 55; and social housing 148–50;
   and social media 188; and warning
   systems 123
co-production 177–87
coastal flood risk 43–57, 72, 94, 142, 147,
   165, 172–3, 177
Cohen, M.D. 80
compensation 20, 23, 30, 34, 137–8, 155–6,
   158, 161
Constructionism/constructivism 2, 8, 166–8,
   170, 177, 183; social constructions 1–16,
   59, 167–8, 170, 177; weak constructivism
   2, 167
consultation 24–5, 70–2, 74
Cook, C. 17
Cook, I.R. 71
Cooper, J.A.G. 161
corruption 47, 50
cost-recovery mechanisms 30–1, 34
crowdsourcing 190, 193–4, 196–7
Cuba 94
Curaçao 4, 43–57
cyclones 94, 124, 165

Dahl, R.A. 44–5, 49, 59
dams 31, 105–6, 133–6, 138, 156, 194
data sharing 119–20, 126
death tolls 58, 83, 95, 116, 119, 121, 125–6,
   133, 137, 145
decentralisation 19, 58, 60–1, 63–6, 81–2,
   138, 190
deforestation 62
Demeritt, D. 6
democracy 47, 81, 183–5
denial 124, 157
developed countries 19, 115, 118, 120
developing countries 6, 20, 43–57, 69–71,
   73–4, 76, 115–30, 132
development 3, 32, 49–50, 52, 106, 122–3,
   160, 165, 168, 173
development banks 75
devolution 19, 58, 60
digital mapping 190
dikes 74, 154–5, 161, 168
diplomacy 75
disaster management agencies 3–4, 85
displaced households 31
Donovan, M.G. 51
Doorn, N. 19
drainage 4, 31, 61–2, 66, 72, 146

dredging 81, 83–4
Duncan, J. 119
dykes 74, 154–5, 161, 168

early warning systems *see* warning systems
ecology 73, 103, 108–10; and the
    Endangered Species Acts 108
ecosystems 18, 104–5, 109
egalitarianism 142, 144–5, 149, 157–8, 160
El Salvador 125
elites 51, 58–9, 61–2, 66, 160
embankments 1, 74
emergency intentional flooding 9, 153–64
Emerson, R.M. 44, 50
engineering 4, 7, 9, 11, 61, 63, 72, 74, 103,
    105–6, 180
England 4, 144, 146, 178, 181, 183
entrepreneurs 80–1, 84, 87, 93, 99
Environment Agency 4, 144, 188
environmental justice (EJ) 154, 166, 168–71
Environmental Protection Agency (EPA)
    95–7, 177
environmentalism 104, 107–12, 158
epistemology 6, 59
Epple, K. 21
escalator effect 168
ethics 74, 76, 149, 182
ethnicity 121, 125, 172, 183
European Union (EU) 17–18, 21–2, 24, 65
evacuations 11, 106, 118, 124–6, 132, 137,
    158
Evans, M. 70
exclusion 2, 7, 120–3, 127, 182
experts 7, 9, 25, 47, 69–71, 73, 75–6, 80,
    120, 168, 181–5
extreme weather events 31, 91–2, 96–7, 99

Facebook 190–1, 193–4
fake news 196
Farcomeni, A. 58–68
fascism 62
Federal Emergency Management Agency
    (FEMA) 95, 165, 178–81, 191
Feltmate, B. 34–6, 39
finance 30–42; and challenges 31–39
financial crisis 81
First Nations communities 155–6,
    158–9, 161
Fischer, L.W. 165–76
flood damages 32–3, 95, 98, 105, 107; and
    emergency intentional flooding 153,
    155–9, 161–2; and finance 30–42; and
    flood insurance maps 178; and social
    housing 145, 149; and warning systems
    116, 119, 133, 137–8

flood insurance maps 177–87; and
    challenges 179-80; and climate change
    177, 179, 185-6;
Flood Re 39, 145
flood risk management (FRM) 1–16; and
    challenges 1–5, 7, 10–13, 17, 19; and
    climate change adaptation 91–102; and
    coastal flood risk 43–57; and Dutch flood
    policy 69–78; and emergency intentional
    flooding 153–64; and financing flood
    damages 30–42; flood disadvantage 146–
    7; flood-proofing 143–4, 178; flood-safe
    architecture 21; and inherent
    complexities 10–11; and legal geography
    17–29; and recovery 131–40;
flood defences 3–4, 9, 17–18, 39–40, 52–3,
    72, 81, 105–6, 132, 141, 188; and flood-
    proofing 143–4, 178; and flood-safe
    architecture 21;
flood risk communication 8, 20, 132, 136,
    157, 188–201; and flood risk warning
    chains 10, 116–18, 120, 123, 125–7,
    131–2, 136, 138, 157
Flood Risk Management Plans 18, 65;
    flood-proofing 143–4, 178; flood-safe
    architecture 21; and floodplain
    occupation 165–76; and inherent
    complexities 10–11; and legal geography
    17–29; and multi-hazard early warning
    systems checklist 127; and multiple
    streams model 79–90; and policy belief
    change 103–14; and power shifts 58–68;
    and recovery 131–40; and social housing
    141–52; and social justice 141–64; and
    social media 188–201; and Superstorm
    Sandy 91–102; and warning systems
    115–40
Flood risk management plans 3-4, 11, 17-9,
    24-6, 65, 82-3, 95-9, and spatial planning
    9-10, 20-1, 35, 51, 65–6, 80–3, 85–6,
    116–17, 141; and urban planning 47,
    49, 63
flood warning systems *see warning systems*
Flood risk maps 24-6, 35, 45; and Risk
    Mapping, Assessment and Planning (Risk
    MAP) program 178, 180–2, 184, 186
floodplains 3, 6, 20–1, 24, 31; and
    emergency intentional flooding 159–60;
    and finance 34; and flood insurance maps
    177–81; and floodplain occupation
    165–76; and Multiple Streams model 80;
    and policy mobilisation 72; and social
    housing 142; and warning systems 123–4,
    133, 137–8
Floods Directive (FD) 17–18, 24, 65

Floods and Water Act 145
Florida 172–4
forecasting 1, 4, 65, 92, 106, 116–20,
    126–7, 132, 134–6; and detection
    116–20, 127, 132, 134–6, 190–1, 198
forums 64, 66, 105, 109–10, 190
Foucault, M. 2–3, 6
frameworks of reality 10–12
France 22, 38
Fuldauer, L. 43–57

G8 30, 38–9
game theory 60
Ganges River 120, 126
Garrick, D. 17
gender 125
geomorphology 153
geotagging 192–3, 197
Germany 5, 10, 17–29, 38–9, 127; East
    Germany 18, 21–2; Elbe River 22–3;
    German National Water Law (WHG)
    18–21, 24–5

Ghana 123, 125
Global Positioning Systems (GPS) 192
Goldman, C. 91–102
Google Maps 190, 194
governance 2, 10, 13, 18–19, 51, 59, 62,
    64–5, 74, 178, 186
greenhouse gas emissions 92, 149
gross domestic product (GDP) 156, 159,
    162
Gulf Coast 172
Gutteling, J.M. 8
Gwimbi, P. 138

Haiti 95
Handmer, J. 117
Helco, H. 94
heuristics 167, 170
Housing Associations 143
Hungary 104, 109
hurricanes 1, 43, 69, 75, 94, 96, 98, 165,
    172, 177–9, 194, 196
hybridity 45, 50, 53, 73, 127
Hydrogeological Setting Plan 64
Hydrological Research Center 126
hydrology 5, 19, 111, 116, 118–20, 126,
    134–5, 154, 160, 181
Hyogo Framework 116, 133

ice jams 31–2
illiteracy 122, 197
India 115, 120, 122, 125; and the Karnali
    River 121, 126

Indigenous communities 10, 123, 155,
    158–9, 161 see also First Nations
    communities
Indonesia 73, 79–90, 124; and Jakarta
    Emergency Management Agency 194;
    and the National Disaster Management
    Agency 85
Indus River 116, 119–20, 122, 126; and the
    Indus Water Treaty 120
information 5–8, 10, 21, 25, 35; and
    coastal floods 46; and dissemination
    7, 25, 71, 104, 111, 116–17,
    120–1, 123, 125, 133, 136, 138,
    189–92, 194; and emergency intentional
    flooding 157; and flood insurance maps
    178, 180–1; and floodplain occupation
    167, 170; and policy belief change 104;
    and policy evolution 92; and policy
    mobilisation 70; and social media
    188–98; and warning systems 116, 118,
    122, 124–5, 131–8
innovation 72–3, 97, 196–7
insurance 1, 3, 7, 9–10, 20–4; and
    The Association of British Insurers
    (ABI) 145, 147; and emergency
    intentional flooding 159–60; and
    finance 30–40; and floodplain
    occupation 173; and Insurance
    Bureau of Canada (IBC) 32, 40; and legal
    geography 26; and mapping 177–87; and
    policy evolution 95, 97; and social
    housing 141, 145–9; and the US
    National Flood Insurance Program
    (NFIP) 7, 38, 168, 173, 177–87; and
    premiums 22, 34–7, 39–40, 145, 173,
    179–80
integrated water management (IWM) 60,
    62–4, 66, 109, 111–12, 118
interdisciplinarity 74, 117
Intergovernmental Panel on Climate
    Change (IPCC) 91–2
intersectionality 20
inundation 74, 81–3, 134, 153, 155–62, 165
investment 4, 34, 50, 132, 142
Italy 38–9, 58–68; and Venice 63;

Jakarta see Indonesia
Jamaica 94
Japan 38–9
Jasanoff, S. 6
Jenkins-Smith, H. 105, 109
John, P. 88
Johnson, C.L. 9, 19, 85, 92, 142
Jones, B.D. 93
Jung, J. 169

Kadzatsa, M. 138
Kahneman, D. 167
Kamala River 126
Karnali River 121, 126
Kates, R. 167
Kearnes, M. 185
Kellog, S. 96
Kenya 121
King, D. 141–52
Kingdon, J.W. 79–80, 85–8
Knight, C.A. 103–14
knowledge 3, 5–8, 11, 24, 30; and
   coastal floods 47, 52; and flood
   insurance maps 177–8, 180–6; and
   floodplain occupation 166; and policy
   mobilisation 69–74, 76; and social
   housing 149–50; and social media
   190, 193; and warning systems
   116–27, 133
knowledge-deficit model 185

Lamond, J. 38
land use 3, 20–1, 35, 50–1, 74, 154, 171,
   179; and land use planning 35, 51; *see also*
   Spatial Planning
landslides 62, 64
Landström, C. 7, 25
Laos 119
lead times 115–16, 118, 120, 136
leadership 9, 51, 97–8, 110–11, 121, 126,
   137, 160
Lee, D. 169
legal geography 17–29
legislation 3, 17–29, 51, 58, 61–2; and flood
   insurance maps 177, 179; and Multiple
   Streams model 80, 85–6; and policy
   belief change 103–4, 107–9; and policy
   evolution 94; and power shifts 64–5; and
   social housing 143; and warning systems
   120, 136
levees 1, 31, 105, 153–4, 168
lobbying 62, 94, 96, 107, 147, 160, 162
locations of control 117, 119–21, 127
Locke, J. 59
looting 124
Lukes, S. 44, 50

Maas, T. 69–78
McCann, E. 70
McGuire, C.J. 168, 173
McKenna, J. 161
McLindin, M. 115–30
Malaysia 168–9
Maldonado, A. 172
mangroves 47, 53

mapping tools 24–5, 32, 35, 39–40, 43–8;
   and coastal floods 53, 55; and flood
   insurance maps 177–87; and power shifts
   65; and social media 190–4; and warning
   systems 126, 131–2, 134–5
March, J.G. 80
Marcus, F. 110
Marks, D. 159–61
Marsh, D. 70
Matthews, D. 44
media 32, 75, 105, 108, 133, 137, 168, 190,
   194; *see also Social Media*
Mekong River 119, 124; and the Mekong
   River Commission 119
meteorology 1, 3, 47, 49, 51–2, 73, 116–19,
   126, 134
Mileti, D. 123
Mill, J.S. 154
Miller, G. 108
minority move-in 171, 174
Minton, J. 110–11
Mississippi River 161
mitigation 13, 32, 34–5, 40, 59; definitions
   92; and emergency intentional flooding
   157; and flood insurance maps 180; and
   floodplain occupation 168, 171; and
   policy evolution 95, 97; and power shifts
   61–2, 64; and warning systems 124
mobility 69–71, 73–6, 143, 149, 157, 169
modelling 5–6, 9, 73–5, 80–1, 86–7; and
   emergency intentional flooding 157; and
   flood insurance maps 178, 180–5; and
   warning systems 132, 134–5, 138
monsoons 156
Montgomery, M.C. 169, 172
moral hazard 34, 159, 179
Mugabe, R. 10
Mukorsi River 133
multi-level power dynamics mapping
   (MLPDM) 45–6, 49–50
multi-level stakeholder mapping (MLSM)
   45–6, 48, 53, 55
Multiple Streams (MS) model 79–90; and
   problem stream 80–2, 84, 86, 93, 99; and
   windows of opportunity (WOP) 52, 63,
   85–7, 93–4, 99–100
Mussolini, B. 62
Muter, A. 153–64
Myanmar 122
Mycoo, M. 51

negotiation theory 60–1, 65
Nepal 120–1, 125–7
Netherlands 6, 8, 51, 69–78, 178; and the
   Deltaprogramme 72; and the Netherlands

Water Partnership (NWP) 75; and the
  poldermodel 72
Newig, J. 24
Nielson, I. 96–7
Niger 126
Nigeria 71, 125
Nile River 119
non-governmental organisations (NGOs)
  3–4, 46–7, 49–51, 53, 55, 73, 97–9,
  134–6, 196; and the Red Crescent 3,
  137; and the Red Cross 3, 137

non-human actors 44–5, 48, 55
normativity 2, 5–8, 12–13, 24, 74, 104, 125,
  145; scientific normativity 5–6; socio-
  political normativity 5–7; technological
  normativity 5–6

Obama, B. 95, 97–8
Octavianti, T. 79–90
Oels, A. 19
Olsen, J.P. 80
ontology 11–12, 44, 59–60
Oulahen, G. 30, 36–7
overland flooding 32–7, 40

Pakistan 116, 119–20, 122, 125
Pardoe, J. 168–9
Paris Climate Accord 95, 98
Parker, D.J. 116–17, 168
Paul, S. 123
Penning-Rowsell, E.C. 1–16, 19, 38, 59,
  141–52
Perda 81
Pfohl, S. 2
Philippines 127, 192
Pidgeon, N. 9, 19
Plate, E. 123
pluralism 45, 55, 59, 66
policy 79–90; and climate change adaptation
  91–102; and coastal flood risk 43–57;
  Dutch flood policy 69–78; and
  emergency intentional flooding 153–64;
  and financing flood damages 30–42; and
  flood insurance maps 177–87; and flood
  risk communication 188–201; and
  floodplain occupation 165–76; and legal
  geography 17–29; and multiple streams
  model 79–90; policy belief change 103–
  14; policy core beliefs 104–10; policy
  evolution 91–102; policy mobilities 69–
  78; policy models 92, 94, 99, 101, 103;
  policy stream 80–2, 86–7, 93, 99; policy
  subsystems 94, 103–4, 109; policy
  tourism 71, 75; and power shifts 58–68;

and recovery 131–40; and social
  constructions 1–16; and social housing
  141–52; and social justice 153–64; and
  social media 188–201; and Superstorm
  Sandy 91–102; and warning systems
  115–40
policy-oriented learning (POL) 104–5,
  109–11
political economy 59
political science 69–70, 92
politics 1, 6–7, 9–10, 12, 17–19;
  and coastal floods 47, 50–3, 55;
  and emergency intentional flooding
  160–2; and flood insurance maps
  177–8, 180, 182, 185; and floodplain
  occupation 168; and policy evolution
  96–7; and policy mobilisation 70, 75;
  politics stream 80–2, 84, 86–7, 93–4,
  99; and power shifts 58–62, 66;
  and social housing 143, 145;
  and warning systems 120, 122, 124,
  133, 138
pollution 171, 173–4
positivism 5–7, 12
post-structuralism 59
poverty 115, 123, 158, 192
power 1–7, 12–13, 18, 24, 43–54, 58–60,
  66, 167, 182–5; and climate change 43,
  51, 55; and multi-level power dynamics
  mapping (MLPDM) 45–6, 49–50
Priest, S.J. 9
processual approach 60
property rights 20, 23
psychology 8, 105, 112, 122–4, 127, 158,
  166, 169–70, 198
public good 20, 23, 25
public opinion 35, 37, 183
public participation 20, 24–5, 36–7, 74, 107,
  133, 161, 177–8, 180–6, 188, 194, 198
Puerto Rico 98
Punctuated Equilibrium Theory (PET)
  93–4, 99–100

Rahman, M. 121
rational choice theory 60
rationalism 11, 166–7, 170
Rawls, J. 142, 144–5, 149, 154, 157–61
realism 1–16
recovery 131–40
refugees 172
Reilly, A. 177–87
religion 124, 160
renewable energy 98
repairs 31, 109, 126, 143, 146–8
reservoirs 84, 105, 154, 156

resilience 19, 22–3, 62, 66, 72; and
emergency intentional flooding 154, 158,
160; and policy evolution 92, 96–8; and
policy mobilisation 74; and social housing
143–4, 146, 149; and social media 188,
191, 194, 198; and warning systems 115,
118, 123
response 1, 8, 12, 25, 51; and coastal floods
54; and emergency intentional flooding
153; and Multiple Streams model 79–80,
84–5; and policy belief change 103–14;
and policy evolution 91, 97, 99–100; and
policy mobilisation 72; and power shifts
65; and social media 190, 194–6, 198;
and warning systems 116–17, 121–7,
131–4, 136–8
retention ponds 84
revolving door framework 91–102
riparian flood risk 119–20, 165, 169,
177, 181
risk 1–16; coastal flood risk 43–57; and
Dutch flood policy 69–78; and
emergency intentional flooding 153–64;
and financing flood damages 30–42; and
flood risk communication 188–201; and floodplain
occupation 165–76; and legal geography
17–29; and multiple streams model 79–
90; and policy belief change 103–14; and
power shifts 58–68; and recovery 131–40;
risk perception 7, 9, 166–8, 170–1, 173–
4, 189, 198; and social housing 141–52;
and social justice 153–64; and social
media 188–201; and Superstorm Sandy
91–102; and warning systems 115–40
Risk Mapping, Assessment and Planning
(Risk MAP) program 178, 180–2, 184,
186
River Basin Districts 18
River Basin Management Plans 18
Rohinhya communities 122
Room for the River 72–3
Rose, N. 19
Rousseau, J.-J. 59
Routray, J. 123
Russia 38–9

Sabatier, P.A. 79, 105, 109
safety 4, 21, 23, 47, 72, 124–5, 168, 191,
196
sandbags 144, 156
Sandink, D. 33, 37, 39
satellites 126, 135, 157
Sayers, P. 146–7
scalar processes 58–60, 63–4, 66, 71, 161

Schäfer, R. 21
Schwarze, R. 23
science and technology studies 6, 70, 177
scientific normativity 5–6
Scotland 147–9
sea-level rise 10, 84
sea-surges 43, 45
seawalls 168
secondary beliefs 104
Self, J. 2, 59
Senegal 118–19
sewers 32
short-termism 161
slums 124
Snow, L. 110–11
social capital 149
social constructions see also constructionism/
constructivism 1–16, 59, 167–8, 170, 177
social housing 141–52; and challenges 149;
and climate change 148–50; and Decent
Homes 143–4, 146, 149
social justice 9–10, 12, 25, 74, 141–50,
153–62
social media (SM) 5, 7, 121, 188–201;
and blogs 190, 194; and challenges 193;
and climate change 188; and digital
mapping 190; and Facebook 190–1,
193–4; and Flickr 190; and hashtags
192, 197; and Skype 46; and
Twitter 7, 190–4, 196; Weibo
190, 194
social network analysis 44
sociology 69–70, 167
soil erosion 62
South Africa 71, 132; South African
Weather Service 132
Southern Asia 125
Sova, C.A. 46, 48
sovereignty 119
spatial planning 20–1, 25–6, 35, 65–6, 80–3,
85–6, 116–17, 141; Indonesian Spatial
Planning Act 85, 87; see also Land Use
Planning
Special Flood Hazard Areas 178
squatters 84
stakeholders 5, 7, 9, 17, 20; and coastal
floods 43–8, 50–1, 53, 55; and
emergency intentional flooding 161–2;
and finance 31, 33–6; and legal
geography 24; and multi-level
stakeholder mapping (MLSM) 45–6, 48,
53, 55; and Multiple Streams model 84;
and policy belief change 107, 110–11;
and policy evolution 92–4, 98; and policy
mobilisation 74; and power shifts 60,

63–4, 66; and social media 188; and warning systems 131, 135–6, 138
states of emergency 65
Stengers, I. 7, 183
strategic approach 60
Sub-Saharan Africa 118
subsidence 84, 165
Suharto, H.M. 81–2
Sultana, P. 69
Superstorm Sandy 69, 75, 91–102, 179, 191, 194
sustainability 4, 35, 48, 52–3, 55, 109–10, 116, 141, 149; Sustainable Development Goals 116

Sustainable Development Goals 116

Taliban 122
Tarlock, D. 20
taxation 31–2, 36, 39–40, 47, 142, 158, 162
technocrats 74, 107, 116
Temenos, C. 70
temporality 25
Terpstra, T. 8
Tevera, A. 131–40
Thailand 10, 119, 153–4, 156–7, 159–62, 194; and the Chao Phraya River 153, 156–7, 159, 162
Thaler, T. 159
Thislethwaite, J. 34–6, 39
tourism 4, 43, 45–7, 49–53, 55, 71, 75, 156, 172
transboundary flooding 17, 118–19, 125
Trump, D. 95
trust 1, 5–10, 12, 120, 124, 149, 185, 189, 193–8
truth 1, 13
tsunamis 85
Tversky, A. 167
Twitter 7, 190–4, 196
typhoons 192

UNICEF 134, 137
United Kingdom (UK) 4, 9–10, 38–9, 69, 116, 118, 141–52, 169, 178, 188; and the Environment Agency 4, 144, 188; and the Hull River 146
United Nations (UN) 4, 47, 50, 116, 133
United States (US) 7, 17, 23, 38, 69; Army Corps of Engineers 4, 105–7; and the Cleanup and Revolving Loan Fund 97; and climate change 95–96; and Congress 105, 108, 177–80; and the Environmental Protection Agency (EPA) 95–7, 177; flood insurance maps 177–87;

and floodplain occupation 165, 168, 172, 174; and the Miami coast 172–4; and Multiple Streams model 79; policy evolution 92–3, 95–6, 98–9; and social media 191, 196; and warning systems 126
urbanisation 10, 31, 79, 106, 143, 160
utilitarianism 26, 142, 145, 157, 159–62

Venice see Italy
vernacular knowledge 116–17, 183–4
Vietnam 115, 119, 121
Vogt, M. 171
Volta River 125
voluntariness 169–74
vulnerability 8–11, 20, 23, 26, 36; and coastal floods 43, 48, 53; and emergency intentional flooding 157–60; and flood insurance maps 177, 179–80; and floodplain occupation 165, 172–3; and Multiple Streams model 79; and policy belief change 109; and policy evolution 91–2, 96, 99; policy mobilisation 74; and social housing 142–3, 146–50; and social media 188, 192, 197; and warning systems 115, 117, 119–20, 122, 125

Wagner, C. 127
Wagner, G.G. 23
Wahl, T. 172
Ward, K. 70–1
warning systems 1, 3, 6, 10, 49 115–38; and challenges 115–30, 132, 136, 138; and climate change 123; and dissemination 116–7. 120–21 and Multiple Streams model 85; and power shifts 65; and recovery 131–40; and SMS warnings 122, 125; and social housing 141, 144; and social media 189–90, 193, 196–7
water boards 72
Water Framework Directive (WFD) 17–18, 24
weak constructivism 2, 167
Weber, M. 2
weirs 105
Whatmore, S.J. 184
wildlife 103, 106–8
willingness-to-pay studies 35, 39
Willis, K.F. 169
Wilson, P. 110
windows of opportunity (WOP) 52, 63, 85–7, 93–4, 99–100; see also Multiple Streams model
Wood, A. 71, 75
World Bank 4, 47, 50

World Meteorological Organisation (WMO) 126–7
Wu, W. 188–201
Wynne, B. 25

Yingluck Shinawatra 159

Zambezi River 119
Zhang, Y. 171

Zia, A. 127
Zimbabwe 10, 131–40; and the Mukorsi River 133; Tokwe River 133; Tokwe-Mukorsi Dam 133–8; Zimbabwe National Water Authority (ZINWA) 134–5, 137–8Zirecho, S. 137
zoning 3, 21, 31, 34, 36–7, 48, 59, 157, 173, 179; *see also Spatial Planning*